P9-DCM-706

CONTENTS

1
Introducing the
Greek Islands

With over 100 inhabited islands – and thousands more uninhabited isles – Greece's seas tempt you with the promise of escape to a world of whitewashed lanes, olive groves, azure seas and endless sunshine.

The reality is a little different. Many of the islands (notably **Corfu**, **Kos**, **Zakynthos**, **Rhodes** and **Crete**) have surrendered to package tourism, and even the remotest islands receive their share of summer visitors. Indeed, the smallest islands can seem even more crowded than their larger neighbours, with smaller beaches to be shared. But even on the busiest islands there are quieter villages and emptier beaches off the beaten track, and Greece's archipelagos offer everything from busy resorts with garish nightlife to miniature harbours where a quiet after-dinner drink at a rickety café table is the summit of excitement.

Greece's ancient culture has left its mark on all of western civilization through its philosophy, art, architecture, and most of all its first experiments with democracy and urban society. The sea has always been central to this culture: a Greek proverb says 'the land divides, but the sea connects', and far from being cut off from the heart of things, the islands were for centuries more prosperous and populous than much of the mainland.

Most visitors come to Greece during July and August, the hottest time of year. Walkers and nature-lovers will prefer April and May, and October is great for those who want peace and quiet, while the sun and

ALBANIA
TURKEY
●ATHENS
MEDITERRANEAN SEA

TOP ATTRACTIONS

*** Athens: Acropolis and Parthenon – the most famous temple in Greece.
*** Corfu: the magnificent Venetian town.
** Delos: the lions and other antiquities of the centre of worship in ancient Greece.
** Rhodes: Street of the Knights and the wonderful Medieval architecture.
** Skiathos: renowned for the best beaches in Greece.
** Santorini (Thira): Akrotiri and the Minoan town.
* Crete: Knossos – painted remains of a Minoan city.

Opposite: *Island churches are distinctive with domes and whitewashed walls.*

sea are still warm. Many tourist facilities close down
from late October until Easter, and from November to
March the weather is often wet, cold and windy, with
snow on the mountains of islands such as Crete,
Samothraki, Kefalonia or Ikaria.

Larger islands are linked with Athens and Thessaloniki (Greece's second city) by regular domestic flights
and ferries which operate all year round, and frequent
charter flights operate weekly and fortnightly from
British and European cities to around a dozen holiday
islands between early April and mid- to late October.

THE LAND

Although the Greek islands tend to be grouped for
administrative and tourist convenience, each island
retains a strong identity, particularly as far as proud
islanders are concerned. Islands in the west (**Ionian**)
and northeast (**Aegean**) lie near huge landmasses,
receive more winter rain and appear lush and green in
comparison with the dry **Cyclades** and their wild, bare
landscapes of limestone and sandstone rocks. A few
islands – **Santorini** (Thira) is the best known – possess
a lunar landscape of dark volcanic rocks. The higher
mountains – more than 1500m (4922ft) – are confined
to the larger islands (**Crete, Kefalonia,**) but most of
the islands are hilly, some of the smaller ones being
little more than the tops of mountains, left when the
Mediterranean basin was flooded in the distant past.

Many islands remained forested until around 8000
years ago – the dawn of 'civilization': today each island
is a mosaic of habitats shaped by human intervention.
Yet, the scrub-covered hillsides and poor soils which
feature so much in the Greek-speaking world are
botanical treasure houses from which over 6000
species of flowering plants have been recorded.

Seas and Shores

Greek beaches are officially public although some
beach hotels act as if their strip is not. On some islands
beaches are designated for nude bathing – in practice,

away from towns many visitors bare all. Superb, sandy beaches are found on **Rhodes**, **Naxos**, **Skiathos** and many other islands. There are tiny coves, long stretches of golden sands, some crowded, others barely discovered – something for everyone.

Greece boasts over 14,500km (9000 miles) of coastline but there is little shallow water around the islands, except close to the shoreline. The **Ionian Sea**, the central basin of the

Mediterranean, is well over 3500m (11,484ft) deep in a few places. By comparison, the **Aegean** is a shallower sea where winds create hazardous conditions which can make landing on rocky coasts a nightmare. Sensibly, the sailors of ancient Greece largely abandoned voyages during the winter months. The strong winds are encountered between April and October and during peak holiday times, providing some relief from high temperatures rather than proving a nuisance.

Above: *Mountains provide a backdrop to many Greek villages – nowhere more dramatically than Crete where the White Mountains are snow-capped for much of the year.*

Climate

The **Mediterranean** climate is characterized by mild, damp winters and hot, dry summers separated by short spring and autumn periods. Most **rain** falls in the winter and early spring months (**December to mid-March**). **Snow** occurs regularly only on the higher mountains.

COMPARATIVE CLIMATE CHART	IONIAN ISLES				AEGEAN ISLES				CYCLADES			
	SUM JAN	AUT APR	WIN JUL	SPR OCT	SUM JAN	AUT APR	WIN JUL	SPR OCT	SUM JAN	AUT APR	WIN JUL	SPR OCT
MAX TEMP. °C	10	15	27	19	10	16	27	19	12	17	25	20
MIN TEMP. °C	50	59	81	66	50	61	81	66	54	63	77	68
MAX TEMP. °F	13	16	24	22	11	14	25	21	15	16	24	21
MIN TEMP. °F	55	60	75	71	52	57	77	70	59	61	75	70
HOURS OF SUN	4	7	12	7	4	7	12	7	5	7	12	7
RAINFALL in	4.5	2.5	0	4	3	2	1	3	2.5	1	0	1.5
RAINFALL mm	111	62	6	105	71	48	19	73	66	18	1	36

A Salty Tale

The **Mediterranean** sea loses far more water by evaporation than it gains from all the rivers which drain into it: the difference is replaced by a steady inflow of nutrient-rich sea water from the **Atlantic**. Thus, when visitors say it is easier to swim in the Mediterranean they are right: evaporation increases the salt content and this creates a greater degree of buoyancy.

THE MELTEMI

High temperatures in **July** and **August** are made tolerable throughout the islands by the *meltemi*. This wind begins as a breeze at dawn, rises to a crescendo around midday and falls almost imperceptibly towards evening. Sailors in north-facing bays need to set sail early or face being trapped until the wind dies down.

Opposite: *Bright pink Judas trees in blossom.*
Below: *Spring brings a rush of colour to the fields such as those near Aradhena in Crete.*

The climate of individual islands shows variations: the northeastern Aegean islands have a slightly cooler, wetter climate than islands in the south while the islands to the west are noticeably drier than those in the eastern part.

Winds have long dictated the ease of passage between the islands and the most effective sea routes. Hot summer air brings the ***Etesian*** winds which once filled the sails of Mycenaean ships: boats were hauled ashore in autumn and then re-launched in spring when the lesser ***Prodromes*** began to blow. The ***Etesian*** winds give a clarity and briskness to the Aegean air and wonderfully clear night skies. Ionian skies have a hint of haze about them in summer. In spring and autumn the ***Sirocco*** blows: in its travels it collects moisture increasing humidity (and human lethargy). The ***Gregale*** is a winter wind of moist Atlantic air that brings heavy squalls in the Ionian islands, rapidly darkening clear skies and forcing fishermen close to shore.

Plant Life

The study of Greek plant life began with the philosopher Aristotle in the 4th century BC. One of his brilliant pupils, Theophrastus, wrote on orchids and is regarded as the first true botanist, while Dioscorides wrote his well-known *Materia Medica* on the medicinal uses of plants in the first century AD.

Many of the plants of mainland Greece are also found on the islands. In addition, isolation has given some plants the time to evolve and become quite distinct from their mainland relatives. Extensive **salt marsh** and **sand dune** areas are unusual in the islands, yet sandy soils within a few metres of the sea shore have **yellow horned poppies**, **purple sea stock**, **scarlet poppies** and, in the height of summer, **white sea lilies**. Large cultivated areas are as sterile in Greece as they are elsewhere whereas small, stony fields are a haven of species, forming an astonishingly colourful patchwork of **white chamomile**, **yellow crown daisies**, **scarlet poppies** and splashes of **blue anchusa**.

Phrygana – low, scented scrub – colours the hills in spring when the woody herbs (oregano, thyme and sage)

and **rockroses** (*cistus*) flower. Hillsides are dotted with the thin, dark forms of the **funeral cypress** and enlivened by springtime splashes of pink when the **Judas trees** flower. Phrygana is succeeded by taller bushes and small trees (mastic, strawberry tree and myrtle) and then by open woodlands of **Aleppo pine** *(Pinus bratia)* with an understorey of shrubs. Significant forest areas still occur on some of the larger islands (Chios, Samos, Lesbos). Mt Ainos on Kefalonia (1620m; 5315ft) is thickly wooded with **Greek silver fir** (*Abies cephalonica*). Elsewhere, **Aleppo pine** is the dominant species on lower mountains, giving way to the **black pine** (*Pinus nigra*).

Only Crete has mountains high enough to possess a genuine high-mountain component in its flora: **Cretan crocus** (*Crocus sieberi*), **Cretan chionodoxia** (*Chionodoxia cretica*), mats of a striking, blue **bugloss** (*Anchusa caespitosa*) and the **Ida arum** (*Arum idaeum*) among many others.

Conserving the Natural Heritage

Greece has in recent years become active in conservation even though some of its famous sons were writing about the denudation of hills and dangers of erosion over 2000 years ago. The small but dedicated membership of the **Hellenic Society for the Protection of Nature** (9 Kidathineon Street, Athens) has been a driving force.

Because of the mountainous nature of much of its territory Greece has largely escaped the ravages of intensive

BULBS

Stony fields and scrubby hillsides on lime-rich soils offer an array of bulbous plants which appear with the first rains. The autumn-flowering species: yellow **sternbergias** and pink **colchicums** bring welcome colour to parched hillsides, yet the main explosion occurs in spring time.

In the far southern islands spring begins as early as February with the first rush of **anemones** followed by **grape hyacinths** (*muscari*), **crocuses**, **irisis**, **tulips** and then vast numbers of annual species – the Aegean islands have a distinctly 'eastern' flavour to their bulb flora.

WILD ORCHIDS

Some species of wild orchids, especially the genus *ophrys*, mimic insects in shape, colour and scent in order to attract them for pollination. These orchids draw numerous enthusiasts to the islands each year – favourite destinations are Crete, Corfu, Rhodes, Lesbos and Samos. Crete alone boasts nearly 70 species of wild orchid including the **Cretan bee orchid** (*Ophrys cretica*) and the **hooded helleborine** (*Cephalanthera cucullata*) that are found nowhere else in the world. On Lesbos a few plants of the rare and beautiful **Komper's orchid** (*Comperia comperiana*) grow in the mountains.

The monk seal (*Monachus monachus*) is now Europe's rarest mammal and features prominently on a list of the dozen most endangered animals in the world. Pressures from tourism have driven the seals from the isolated caves, beaches and rocks they once inhabited and fishermen have long persecuted them, seeing them as competing for the diminishing fish populations. Numbers have been severely reduced – current estimates of seals are between 350 and 700. Breeding colonies have been established on tiny islands like **Piperi**, east of **Alonissos**. Females give birth to only one pup every two years and so recovery is bound to be slow.

Below: *Tree frogs, although rather small, make a noise out of all proportion to their size.*

cultivation with the attendant cocktail of pesticides and herbicides. Around the cities and major ports pollution is at danger levels and on some islands over-development for tourism has destroyed many rich habitats. There are 10 areas in Greece designated as national parks (three on the islands) but little in the way of a substructure of officials to care for them (Crete's Samaria Gorge is the exception). On a positive note, environmentalists are pushing for legislation and the seas around the northern Sporades have been designated a national marine park. Efforts are also being made to protect the endangered **monk seal**, **green** and **loggerhead turtles** and **Eleanora's falcon**.

Wildlife

In the Pleistocene period, some of the bigger islands were home to the **dwarf elephant**, **pygmy hippopotamus**, **ibex**, **genet** and **wild boar**. Today, hunting and deforestation have almost driven the last of the larger mammals from the islands, although **badger**, **fox**, **beech marten**, **weasel** and **hare** have survived in isolated areas. The **kri-kri**, or Cretan ibex, has a stronghold in the White Mountains of Crete and on its offshore islands (Dhia and Agios Theodori). Crete also has an endemic animal – the **Cretan spiny mouse** (*Acomys minous*), which is unique to Crete and has spines rather than hairs on its back. It is an inhabitant of rocky hillsides. The most widespread Mediterranean cetacean is the **common dolphin**, often seen in schools 'surfing' the bow waves of ferries. **Bottle-nosed dolphins** are also often spotted.

For the bird-watcher the best times to visit coincide with the spring (Apr to mid-May) and autumn (late Aug to Sep) migrations, though these are never easy to predict. In spring there are vividly coloured **hoopoes**, **rollers** and **bee-eaters** as well as a host of small **warblers**. At lake margins large water birds include herons, **bitterns** and **egrets**. In the White Mountains of Crete look out for the four European species of vulture including the **lammergeier** (bearded vulture) and **blue rock thrush**. **Eleanora's falcons** raise their broods to catch the autumn migration on islands off the north coast and come to Sitia to feed.

Above: *The swallowtail, one of the most colourful of all Greek butterflies.*

From April onwards **butter-flies** appear: **swallowtails** both common and scarce, **southern festoons** (Crete has an endemic festoon) and a yellow, brimstone relative with bright orange patches – the **Cleopatra**. **Hawk moths** are common – especially the tiny whirring forms of the **hummingbird hawk** which flies by day. **Jersey tiger moths** are found on most of the islands but in Rhodes they were responsible for the famous displays at Petaloudes (Butterfly Valley, *see* p. 78).

Praying mantises rest on grasses and shrubs and you may see **cicadas** on branches or in whirring flight, though cicadas are more often heard than seen – in high summer their shrill song can be almost deafening. **Mosquitoes** are pests everywhere. Most bedrooms have plug-in repellents which you can buy in most shops. Locally sold liquid repellents are often less effective – bring a roll-on or gel with a high DEET (di-ethyl toluamide) content from home.

The **blunt-nose viper** (*Vipera lebetina*) is the only snake dangerous to humans; it has a distinctive yellow, horn-like tail. Lizards are seemingly everywhere, from tiny pale geckos to the large **Balkan green lizards** (*Lacerta trilineata*). Tiny, green **tree frogs** (*Hyla arborea*) can be heard at night in the trees: in spring any pond is alive with vocal **marsh frogs** (*Rana ridibunda*), and you might hear the 'plop' of a submerging **striped-necked terrapin** (*Mauremys caspica*).

Island beaches tend not to be rich in shells but close to shore **sea urchins**, colourful **peacock wrasses** and scarlet **soldier fish** can be seen when snorkelling.

TURTLES

Two decades ago some 1500 loggerhead turtles (*Caretta caretta*) nested annually on the sandy beaches of Zakynthos: now their numbers have been more than halved as tourist pressure has driven them from Laganas Bay to other beaches. Power boats and jet skis can cut them, plastic bags choke them and, at night, beach lights or activity prevent them from emerging to lay their eggs. Hatchlings are extremely vulnerable as they crawl to the sea. Luckily a dedicated group of people is fighting to make sure turtles and tourists co-exist.

If you are interested in helping, write to the **Sea Turtle Protection Society of Greece**, PO Box 511, 54 Kifissia, 14510 Greece.

Above: *Delos, a religious and commercial centre in Classical times.*

HISTORY IN BRIEF

Evidence of human activity in Greece stretches as far back as 8500BC to proto-Greek settlers from the eastern **Mediterranean**. The peoples of **Crete** and the **Cyclades** were constructing superb palaces around 3000BC.

Early Cycladic 4500–2000BC

The islands were colonized and small agrarian and fishing communities were established. The first traces of a 'civilization' have been found in the **Cyclades**, where a distinct culture flourished from 3200–2000BC until trading brought in outside influences. Its main legacy is the small marble figurines found in tombs.

Minoan 2000–1500BC

The Bronze-Age Minoans were the first major power in the Aegean – with great palaces on **Crete** at Knossos, Phaestos, Zakros, Malia, Hania and the ancient settlement of Akrotiri on **Santorini**. Agriculture underpinned the **Minoan** economy and enabled them to indulge a passion for art. Minoan civilization literally came tumbling down in 1500BC when Santorini, the nearest of the Cyclades, exploded in a volcanic eruption.

Mycenaean and the Dark Ages 1500–776BC

The Mycenaeans from the Peloponnese also evolved a successful palace-ruled civilization: they succeeded the Minoans and absorbed a great deal of that culture. Essentially a military people, their weapons and funeral artefacts were nevertheless richly decorated with gold and jewels. They built a trading empire across the eastern **Mediterranean** and became embroiled in a 10-year war against Troy. **Dorians**, often regarded as 'barbarians', invaded from the north in 1100BC: palaces fell and trading was interrupted. **Phoenician** traders revitalized those islands which lay on their trade routes: Greek culture began to flourish, city-states arose and the Olympic Games were established in 776BC.

Archaic Period 776–490BC

This was a dynamic and experimental period before art became 'Classical'. Greek architecture borrowed ideas of scale and the use of supporting columns from Egyptian builders but evolved the elegant lines associated with later Classical buildings: statues and carvings, too, began to take on realistic rather than representational proportions. A number of islands came into prominence as powerful island states. **Aegina's** silver coins became a common currency in much of the Mediterranean. **Delos** achieved renown as the shrine to Apollo. **Samos** became a democracy in 650BC, then the tyrant Polykrates assumed power in 550BC and became the first to rule the Aegean since King Minos. The people of **Sifnos** were reputed to be the wealthiest in Greece because of the gold and silver mined there. **Santorini** began to mint its own coinage in the 6th

CULTURE VULTURES

The cultural explosion associated with Classical Greek civilization was centred on Athens: **Aristophanes** wrote his comedies, **Aeschylus**, **Euripedes** and **Sophocles** penned great plays, **Phidias** created sculptures, **Socrates** founded a whole school of philosophical thought later developed by **Plato**. The islands produced their important figures, too. **Hippocrates** (Kos – father of medicine), **Pythagoras** (Samos – mathematician and philosopher) and **Sappho** (Lesbos – poetess).

HISTORICAL CALENDAR

7000–2800BC Neolithic Era.
4000 Settlements on Limnos.
3000–2000 Early Cycladic.
2800–1000 Bronze Age.
2600–2000 Early Minoan culture in Crete.
2000–1700 Middle Minoan –Crete major sea power.
1700–1450 Late Minoan Age.
1600–1150 Mycenaeans rule in Peloponnese: occupy Crete and Rhodes.
1150 Dorian invasion, Ionian settlement of Aegean islands and Asia Minor. Start of the Dark Age in Greece.
500–323 Classical Age, Parthenon finished.
478 Delos made the centre of a Maritime league.
431–404 Athens crippled by Peloponnese war.
378 Second Delian League.
338 Greece conquered by Philip of Macedon.
334–323 Alexander the

Great establishes his empire.
180BC–AD395 Roman Age.
AD395 Byzantine Period.
AD58 St Paul visits Lindos.
824–861 Saracen/Arab occupation of the islands.
1204 Fall of Constantinople – Venetians control islands.
1261 Constantinople taken by Greeks from Latins.
1309 Rhodes becomes base for Knights of St John.
1453 Turkish conquest of Greece begins.
1522 Knights of St John defeated by Ottomans.
1669 Venetians lose Irakliou (Crete) to Turks.
1796 Napoleon captures Venice and Ionian Islands.
1815–64 Ionian Islands under British Rule.
1821–27 Greek War of Independence.
1912–13 Greece gains Crete and northeast Aegean Islands

in Balkan Wars – Dodecanese islands gained by Italy.
1922–23 Greece invades Turkey. Brutal reprisals.
1941 Germany invades Crete.
1948 Dodecanese returned.
1949 Civil war ends.
1967 Colonels' Junta.
1981 First PASOK government, Greece joins EU.
1990–93 Nea Demokratia government under Mitsotakis.
1993–2004 PASOK back in power.
2001 Greece adopts the Euro.
2004 Greece wins Euro 2004 soccer championship; ND re-elected; Athens hosts Olympic Games.
2005 Karollos Papoullas elected president.
2007 Severe forest fires in Attica and islands including Skiathos and Evia.
2007 ND re-elected with reduced majority.

THE CYPRUS QUESTION

The Turks with Britain and Greece were appointed 'guarantor powers' after Cyprus gained independence from Britain in 1964. The Turks had long threatened to invade Cyprus to 'protect' the Turkish minority and the action by the Colonels (see p. 17) provided an ideal excuse.

Greeks were forced from the north – the invasion was brutal and condemned internationally. Since 1974 the island has remained partitioned with the north recognized as a state only by Turkey. Since the invasion many Turkish Cypriots have left the north and its economy is faltering, while that of the south has flourished.

century and its influence extended not only to Crete, Rhodes and Paros but also to Milos. Persian attacks on the islands became reality until they were repelled by the Athenian-led army at **Marathon** (490BC) and in the naval battle of Salamis (480BC). Athens became a powerful maritime force, and because the Aegean islands had been considerably weakened by Persian attacks, they struck a deal with Athens offering to pay money in return for protection. Known as the **Delian League**, the allowance was exploited by Athens which took marble from the islands, while Pericles used the protection money of the Delian League to pay for the construction of the Parthenon.

Other city-states (**Corinth** and **Sparta**) challenged the supremacy of Athens during the Peloponnese War (431–404BC) but the major players all emerged too weakened to claim any dominance.

Classical and Hellenistic 490–180BC

Alexander the Great accomplished Greek unification in 336BC, marking the transition from the Classical to Hellenistic period. Alexander's empire extended to India and Egypt but on his death from fever at the early age of 33, it was fragmented by rivalries among his generals. Art flourished along lines established in the Classical period,

becoming more ornate. **Samos** and **Rhodes** established reputations as centres of learning to rival Athens.

Roman Period 180BC–AD395

Rome first gained a foothold in the islands as an arbiter in disputes. By 31BC the Roman Empire encompassed the whole of Greece, with the islands divided among five provinces: Epirus embraced the Ionian Islands; Asia, the eastern Aegean islands and the Dodecanese; Cyrenica-Creta included Crete with Libya, while Thasos went with Macedonia, and Achaea included Evia along with the Peloponnese.

The Romans laid roads, built aqueducts and refurbished towns. Wealthy Romans took to visiting the islands, even schooling their sons there and carrying away beautiful statues and carvings to adorn their homes. **Pausanias**, a Roman, wrote the first 'travel guide' to Greece in AD150.

Byzantine Period
AD395–AD1453

The islands faced a millennium of attacks from invaders. Their

proximity to the Greek coast made the **Ionian** islands important staging posts: **Corfu** was attacked by Goths and Vandals in the 5th century and then conquered by Normans in the 12th century. After Constantinople (present-day Istanbul) fell, the islands were given to noble Venetian families as fiefdoms: **Venetian** rule held in the face of **Ottoman** attacks until the islands were gained by **France** under Napoleon Bonaparte in 1797.

Major trade routes to Smyrna (present-day Turkish Izmir) and Aleppo (in Syria) lay via Crete and the eastern Aegean islands: the Cyclades were largely neglected and were abandoned or became the lair of pirates. Venetian rule of a kind came after 1204 when **Marco Sanudo** founded the Duchy of the Archipelago on **Naxos**.

Arab pirates originally targeted the **Aegean** islands as well as the **Cyclades**. **Crete's** strategic position made it a target for Saracen invaders. Venice regarded Crete as its most important outpost but after the Ottoman conquest many families nominally converted to Islam. The Colossus of Rhodes, toppled by an earthquake in 225BC, was removed for scrap by Saracen invaders in AD634. After the fourth Crusade the Genoese were granted sole rights to colonize the eastern **Aegean** and trade with Black Sea ports. From 1204 **Rhodes** first came under Frankish and then Venetian rule. The Knights of St John maintained

Above: *The medical centre – the Asklepeion on Kos – was a place of pilgrimage in Classical times.*

Opposite: *The glory of the Minoan period is revealed at Knossos. Controversial archaeologist, Sir Arthur Evans restored and painted the ruins as they might have been.*

Above: *Greek churches such as that at Andri, Crete, are a treasure house of Christian art.*

their headquarters on Rhodes until the Ottoman siege. Curiously, both **Symi** and **Kalimnos** prospered under the Ottomans through their sponge fishing.

Ottoman Rule 1453–1832

Islands closest to the Turkish coast fell first to the Ottomans but the period of acquisition was a lengthy one – from 1522 **(Rhodes)** to 1715 **(Tinos)**.

Russia held control over a dozen or so islands during its war against Turkey from 1770–74. As Turkish might declined there was an upsurge of Greek nationalism: such insurrection inevitably brought reprisals (in the worst of them 25,000 people were massacred in 1822 on Chios).

Independence

Independence came in 1832 but with it Greece only gained control of the **Cyclades**. Britain transferred **Corfu** and the **Ionian** Islands to Greek rule (1860s) and the **Northern Aegean** Islands were regained at the end of World War I. The **Dodecanese** were in Italian hands from 1912 and taken over by Germany in 1943. In 1947 the Dodecanese passed from Allied to Greek control.

In 1909 the Cretan statesman, Eleftherios Venizelos, formed a new Liberal government. From 1912–13 he led an alliance which fought in the Balkan Wars until Greece entered **World War I** on the side of the Allies. In 1920 monarchist factions forced action against Mustafa Kemal or Atatürk in Ankara (Turkey), which proved disastrous for Greece. In 1923 almost 400,000 Muslims (mainly from the islands) left Greece and were 'exchanged' for about 1.3 million Greeks. Those

Greeks and millions of Armenians left in Turkey were systematically and brutally exterminated.

General Ioannis Metaxas, appointed prime minister of Greece in 1936 by King George II, established a state based on fascist lines, but completely opposed Italian or German domination. Greek forces drove Italian invaders from the country in 1940 but the next year German forces conquered the mainland and islands. By the end of 1941 German troops had occupied **Crete** and the islanders were brutally punished for their determined resistance. Nearly 500,000 Greeks starved to death when food was taken for the occupying armies and almost the entire population of Jews was deported to concentration camps.

From 1945–49 Greece became embroiled in a bloody civil war between communist and monarchist forces, with the latter, American-backed, emerging as the victors. On 21 April 1967, a CIA-backed coup d'etat brought the '**Colonels**' to power. Their repressive, fascist regime ended (1974) when they attempted to topple Makarios III in **Cyprus** as a diversionary tactic to recreate a tide of nationalist support. Their efforts failed when Makarios escaped and the Turks were given an excuse to invade **Cyprus** – which they did with unforgivable brutality. It has remained partitioned since.

The Colonels

The Junta's aim was a moral cleansing of Greece – human rights were ignored and censorship became ridiculous. Dissidents and their families were imprisoned and tortured by the secret police: many disappeared. On 17 November 1973, students at the Athens Polytechnic went on strike – the rebellion was quelled with tanks and many died. Papadopoulos, head of the secret police, was arrested. His successor, Ioannides, tried to unite the country on a wave of nationalism by assassinating Makarios and uniting Cyprus with Greece. It failed – the Junta fell and Karamanlis returned from exile in France to form a government.

Below: *Evzone guards on duty at the Tomb of the Unknown Soldier, Athens.*

WOMEN'S RIGHTS

A woman's right to vote was not granted universally until 1956. PASOK fought both its elections (1981 and 1985) with a strong programme for Women's Rights. Extensive changes in family law were achieved in 1983 when equal legal status and property rights were granted to both husband and wife and the dowry custom was prohibited. Women's co-operatives have also been set up to help empower women (*see* p. 19).

Below: *Fertile plains between scrub-covered mountains are intensively cultivated on the islands.*

From 1974 Greece has, once again, enjoyed democratic rule; the monarchy was formally abolished in 1975.

GOVERNMENT AND ECONOMY

Since the restoration of democracy in 1974, Greece has been governed by a 300-seat parliament, with executive powers resting with the prime minister. The president, who is the mainly ceremonial head of state, is also elected every five years. Greek politics is dominated by two main parties, the right-wing Nea Dhimokratia and the left-wing PASOK. ND won the 2004 elections with 45.4% of the vote, ending two decades during which PASOK ruled almost continuously. ND narrowly won the 2007 election with 152 out of 300 parliamentary seats, with PASOK in second place, while KKE (Greek Communist Party) increased its share of the vote from 5.9% to 8.15% and won 22 seats. On the right, the Popular Orthodox Rally party won 10 seats.

Immigration in the 1990s has radically altered the Greek economy. Immigrants – mainly from neighbouring Albania and Serbia – represent some 20% of the workforce but their labour is underpaid and exploited, and with unemployment at 10% there is much resentment of their presence.

Industry and Agriculture

Greece has been a member of the EU since 1981 and adopted the common European currency, the **euro**, in 2001. Membership of the EU has benefited the economy, which saw growth of 3–4% in 2004–05, but joining the euro pushed up prices for Greeks and tourists. Tourism represents some 15% of GDP, industry 22% and agriculture only 7%, but the state now dominates the economy, with 40% of GDP accounted for by the public sector.

On the islands, cen-
turies-old methods of
agriculture are being
abandoned, rather than
modernized, as young
people eschew the hard
life on the land and head
for cities. On the larger
islands, agriculture is the
mainstay of the economy
– **olives** and **citrus** fruit
form the major compo-

nent for home consumption and for export. Locally, crops
like **tomatoes**, **peppers** (and, on Aegina, **pistachio nuts**)
and **fishing** can generate vital income. Manufacturing
industry is essentially low-key (**clothing**, **shoes**, **electrical
goods**), mainland-based and largely for the home market.

There is an omnipresent tension with Turkey (an
issue uniting all parties) not only over the latter's intran-
sigence concerning the 'Cyprus question' (*see* p. 14),
but also its refusal to recognize Greek rights to **oil**
beneath the Aegean seabed. The volatility of the rela-
tionship was shown in early 1996 when a dispute flared
concerning sovereignty of a rocky offshore islet and
troops for both sides were rapidly mobilized. During
the Nato Summit in 1997, mediation attempts of US
Foreign Minister Albright resulted in an agreement
between Simitis and Demirel to resolve existing con-
flicts peacefully.

The Infrastructure

The mainland and larger islands have good **road systems**.
Away from a capital, roads on smaller islands can rapidly
degenerate to a paved (or rough), single-track road. **Public
transport** is often unpredictable; some villages are served
by a bus a day. The **ports** are busy centres for the **ferry
system** – the main commercial link for all the islands. Key
islands in all groups are connected with Athens (Piraeus)
and a few other mainland ports: others can then be
reached by **ferry** and **high-speed catamaran** connections.

Above: *Waterways bustle
with activity.*

WOMEN'S
AGRICULTURAL CO-OP

Traditionally, women in rural
or agricultural areas have
worked physically very hard
on the land: sowing, reaping,
harvesting olives, picking
grapes, tending to livestock
and still rearing children and
doing housework. European
women have largely become
emancipated by industrial
mechanization, but in the
Greek Isles subsistence
farming remains. To facilitate
economic independence and
intellectual acualization of
women, the **Greek Council
for Equality** launched the
Women's Agricultural
Co-operatives in 1985. The
co-operatives attract tourists
who want to experience
Greek farming and village
life and who pay to work
there. The organization and
running is the responsibility
of the women thus enabling
them to acquire managerial
skills and independence.

THE BULL CULTURE

Zeus disguised himself as a white bull and carried off **Europa** who later bore him three sons: **Rhadamanthys**, **Minos** and **Sarpedon**. **Pasiphaë**, wife of **Minos**, fell in love with a bull from the sea sent by **Poseidon** – their unfortunate union resulted in the **Minotaur** (head of a bull on a man's body) whom Minos hid in a labyrinth under his palace. Minos was sent seven maidens and seven youths every nine years from Athens as compensation for the death of his son during an Athenian game. Minos fed the Minotaur the blood of his victims. **Theseus**, son of the king of Athens, wanted to end this cruel practice, so third time round he also went. With the help of **Ariadne**, daughter of Minos, and a long thread, he entered the Labyrinth and slew the Minotaur.

THE PEOPLE

The Greek islands have been part of a greater Hellenic world since Classical times, when Greek colonies spanned the Mediterranean world, from Alexandria in Egypt to Syracuse in Sicily and Marseille in France, as well as trading posts on the coast of Spain and North Africa. Others went north and east to establish Greek cities on the shores of the Black Sea. More recently, during the 19th century, hundreds of thousands of islanders emigrated to North and South America, Australia, and South Africa, as well as to western Europe, so that today there are expatriate Greek communities all over the world. Wherever they go, Greeks have an undying connection with their native land, and many older expats return to their ancestral island to retire. Each island has a distinct character, and there are often strong yet friendly rivalries between neighbouring islands.

Family Life

Island Greeks are very friendly people, and families are very close, even when they span three or four generations and as many continents. Greek children, especially sons, are indulged by parents and grandparents, and visitors with children will finds themselves welcomed more warmly than in any other European country, and solo travellers – male or female – also find a friendly reception in the islands. That said, some female visitors are irritated by the attentions of some young Greek men who assume all foreign women must be looking for a holiday romance. For a foreign male to become involved with a Greek girl still creates opposition even in the most enlightened families.

Religion

Religion has always had a focal position within Greek cultures ever since the first Greeks from Central Asia brought with them the worship of a **mother goddess**. **Christianity** was adopted in Roman times during the rule of the Emperor Constantine. The Old and New Testaments written in Greek in the 4th century AD are in use today by the **Greek Orthodox Church**, to which 98% of the population belongs. Superficially, many Greeks might appear to have moved away from the church – until births, deaths and marriages bring them back. Most people get married in church, sceptics included, although Papandreou legalized civil marriages. Within island communities the role of the church is central and the priest carries great respect within a community. Small **Catholic** populations still exist in the Cyclades – a remnant of almost 300 years of Venetian rule.

Greeks revere their places of worship and, whatever a visitor's views on religion, offence is avoided if they enter modestly dressed. Convention still dictates that widows wear black for several years following the death of a husband or other loved ones – men get away lightly since widowers are only required to wear a black armband for a year.

Tradition dictated that each family built its own chapel – hence the plethora of tiny chapels on the islands. The minarets of derelict mosques (some of which house island museums) are a feature of several Aegean islands, including Lesbos, Chios and Crete, and a tiny Greek Muslim community survives in the old quarter of Rhodes.

Opposite: *For religious festivals traditional dress and customs are preserved.*
Below: *Monks are a familiar aspect of Greek life.*

The Italian influence is unmistakeable in the music of the Ionian isles where much more 'western' melodies are played on guitar and violin. In Crete, Karpathos, Chalki and Kassos the music has a distinctive sound due to the use of the *lyra* – a three-stringed fiddle balanced on the knee and accompanied by the *laoúta*, a lute-like instrument similar to the *oud* of Turkish and Arab music. In Karpathos a primitive bagpipe (*tsamboúna* or *askómandra*) is played and in the Cyclades and most Aegean Islands the *lyra* is replaced by the *violi* (violin).

Myths and Legends

The Greek people retain a strong sense of their past, whether through recorded history or legends. Every island (conscious of the tourist potential) makes great play of its links with the mythical gods and goddesses of the ancient world and tales of their rivalries, affairs and involvement with humans have fascinated people for centuries. The early Greeks were ingenious in incorporating deities into the pantheon of Greek gods and the Romans essentially retained them under different names.

In the beginning … was **Chaos**: from Chaos emerged **Gaia** (Mother Earth) who gave birth to a son, **Uranus** (Ouranos – Sky). The union of mother and her son produced an unholy brood which included the **Titans** and the one-eyed giants, the **Cyclops**. Gaia finally gave birth to a **Golden Race** who lived in a world without trouble or wars: they died out, childless, but their spirits lingered on earth.

Kronos, leader of the Titans, married his sister **Rhea**. It had been prophesied that Kronos would be killed by the hand of one of his own children, so he ate them one by one as they were born. Rhea's sixth child was **Zeus** and she prepared to save him from a similar fate to those of his siblings by giving Kronos a stone to eat. She took Zeus away and he grew to manhood, hidden

Right: *Brightly dressed Corfiot dancers perform traditional routines for visitors.*

from his father. He then returned to fulfil the prophecy by poisoning his father. Kronos coughed up, unharmed and now full-grown, all the children he had eaten – **Pluto** and **Poseidon**, brothers of **Zeus**, and sisters **Hestia**, **Demeter** and **Hera**. They became the main gods of Olympos.

Music

The real music of the Aegean lies behind the bouzouki-dominated façade exemplified by the films *Never on a Sunday* (1959) and *Zorba the Greek* (1964). The melodies have haunting cadences; the rhythms and time signatures are almost hypnotic. Music was an essential element of Classical Greece, with a purity derived from simple five-tone scales without harmonies. Middle Eastern (Ottoman and Arab) and Byzantine music later developed this. For the Greeks music has never been something separate from the people and their emotions. Greek rhythms, alien to lovers of mainstream classical or, indeed, rock music, are certainly familiar to the jazz fan or reader of Homeric verse: 5/8, 7/8, 9/8 and 11/8 timings are intuitive to a Greek musician.

Among popular musicians, **Nana Mouskouri**, **Demis Roussos** and **John Vangelis** have all won considerable success on the international arena.

Dance

At festivals and weddings the dancing is very different from the contrived displays. The secret is *kéfi*, the Greek equivalent of 'soul', where the spirit moves the dancer and the improvised steps provide the lead for others.

In Crete, *pedektó* are very energetic dances with rapid steps while *sírto* embraces slower, almost shuffling dances. The national dance (*kalamatianó*) is a 12-step *sírto* in which a leader improvises and the rest of the dancers – hands joined and held at shoulder level – follow the leader's steps.

Zorba's dance derives from the *hasápiko* (called the butcher's dance) while *syrtáki*, usually performed by two or three men, is another energetic dance.

THE GODS – A WHO'S WHO

GREEK • ROMAN

Aphrodite • Venus
love, beauty, mother of Eros

Apollo • Apollo
poetry, music and prophecy

Ares • Mars
war, son of Zeus and Hera

Artemis • Diana
fertility, goddess of moon, famous huntress

Athene • Minerva
wisdom and war

Demeter • Ceres
harvest and agriculture

Dionysos • Bacchus
wine and fertile crops

Hephaestos • Vulcan
heavenly blacksmith

Hera • Juno
wife of Zeus

Hermes • Mercury
god of physicians, traders and thieves, messenger

Pan • Faunus
part man part goat, god of woods, flocks and shepherds

Persephone • Proserpina
queen of the Underworld

Poseidon • Neptune
sea, brother of Zeus

Zeus • Jupiter
Overlord of Olympos, sky

Right: *Greek statues, even when headless such as these at Delos, have an elegance of line.*

Writers and Poets

Many of the Greek myths can be traced back to the Mycenaean period and even earlier, thanks to **Homer's** *Iliad* and *Odyssey* in which he wrote down the stories belonging to an oral tradition already then approximately 500 years old.

The Homeric tradition influenced countless thousands in Greece and the west as part of a 'Classical education'. The comedies of **Aristophanes** have a mixture of cleverness and unashamed vulgarity which certainly appeals to modern audiences. And Shakespeare drew heavily on this classical legacy with numerous allusions to Greek history and mythology.

Modern Greek authors have gained renown but none more than the Cretan **Nikos Kazantzakis** who penned *Zorba the Greek, Christ Re-crucified, The Fratricides, Report to Greco* and *Freedom or Death.*

Poetry is alive and well in modern Greece with a tradition of intense and dynamic work. Remarkably the country can boast two recent Nobel laureates: **George Seferis** and **Odysseus Elytis.**

Crafts and Customs

Little remains outside museums of the glorious statuary and architecture of the Classical era, but the skills of Christian Byzantium linger on – Greek icon painters, working with traditional materials, are among the world's greatest, though the formal style of painting allows for little individual creativity.

More homespun skills include embroidery, weaving, lacemaking and leatherwork, and fine examples of hand-made clothes, linen, footwear and accessories can be found on many islands. Fine silver jewellery, often using semi-precious stones from the islands, can also be found on more up-market isles such as Santorini and Mykonos.

Sports and Recreation

Basketball is the national sporting obsession, and every island village has its basketball courts. **Football** is another passion and its popularity has grown even more with Greece's Euro 2004 championship victory. Most Greek major-league teams are mainland-based, but islanders follow their fortunes enthusiastically on café TV screens. The Greek national team qualified first in its group for Euro 2008, earning the right to defend its Euro-soccer crown.

Left: *Greek jewellery, often made to traditional designs, is obtainable on all larger islands.*

REMBÉTIKA

Rembétika – reputedly the bouzouki music of dens of iniquity where the women were fast and the smell of hashish hung on the air – began as music of the dispossessed. Prevalent themes were of death, oppression and drug addiction – Greek 'blues'. Much of the music was improvised and played on the Turkish *baglamá* – a forerunner of the bouzouki. The Junta gave an unintended boost to Rembétika by banning it – young people examined the forbidden fruit and recognized the scent of rebellion.

OUT OF SEASON

Henry Miller, Lawrence Durrell and many other writers have travelled to the islands to complete a 'great work'. Whereas larger islands such as Crete and Rhodes can cater for tourists all the year round, small islands literally close shop during the winter months. But if you stay at this time and manage a few words of Greek you are welcomed as a traveller rather than tourist – Greeks are marvellous at spotting the difference.

FISHING

The effects of the inflow of water through the Strait of Gibraltar – with a higher nutrient level – tend to make fishing catches higher in the Ionian islands than the Aegean. In spring and summer, days can be almost windless around the western islands allowing fishermen to make susbstantial catches of **swordfish**, **tunny**, both red and grey **mullet**, **sea bass**, **squid**, **octopus** and the Mediterranean **lobster** (*astakós*). Unfortunately, over-fishing has reduced catches in the past decade, making fish comparatively expensive.

Below: *Traditional cafés abound on islands. Visitors are welcome to sit down and watch the world go by over a cup of coffee.*

In schools, **handball** and **basketball** are popular sports and physical exercise is an important part of the curriculum. For visitors as well as for increasing numbers of Greeks one attraction is the range of water sports from **snorkelling** and **diving** to **water-skiing**, **windsurfing** and even **parascending**. The rich have long continued the national love affair with boats through **private yachts** and **powerboats**. For visitors facilities are always available at the islands' larger resorts and up-market hotels.

The *kafeneíon*, that male-dominated feature of Greek life, is an important part of island culture – here, men sit solving the problems of the day, drinking endless cups of coffee to a background clicking of their **worry beads** or the pieces on a *tavoli* (backgammon) board. Traditional Greek coffee is a finely ground mocha boiled in a pot and served in a small cup. *Kafés glykós* is for those with a sweet tooth, *métrios* is medium and *skétos* unsweetened. A glass of cold water is served with coffee; beware – the coffee is unfiltered and the grounds are in the bottom of the cup. Cafés and bars have until recently been an exclusively male preserve but a younger generation of Greek women now drink with the boys. Coffee in all its forms is the beverage of choice and drunkenness is deeply frowned on.

Food

To sample really good Greek food, try waiting and watching where the local Greeks go.

Estiatórios often serve a variety of casserole dishes: *moussakas* (originally lamb with plenty of aubergine – not the beef and mashed potato tourist version) or *pastíccio* (similar, with pasta rather than potato) – all selected by going into the kitchen and choosing from the pots of food. **Tavernas** range from the grass-roofed beach variety to the specialist **Psarotavérna** (fish) and **Psistaría** (spit-roast meats). They serve a selection of hors d'oeuvres (*mezédhes*) with a main course of fish or meat (grilled or fried), *patátes* (chips) and the ubiquitous Greek salad (*khoriátiki saláta*) of coarse-chopped tomatoes, cucumber, onion and peppers topped with *féta* cheese, olives and a dribble of olive oil.

Starters include *taramasaláta* (made with smoked cod's roe), *tzatzíki* (yoghurt with garlic, cucumber and mint), *eliés* (green or black olives – try the excellent *kalamáta* variety with a pointed end), *rossikí saláta* (potato salad with mayonnaise), *yigándes* (haricot beans in tomato sauce or vinaigrette), *kolokithákia* (deep-fried courgette) served with *skordhaliá* (a simple

Above: *For Greeks eating is a leisurely activity with numerous small dishes, friends and conversation.*

Above: *Fish, although expensive, is a favourite with visitors. Smaller island restaurants catch their own – you dine on the freshest fish imaginable.*

WHERE TO EAT

On the more popular islands, fast-food outlets are well established. Although mainly drinking establishments, ouzeries also serve snacks (*mezédhes*); street stands called giros serve sandwiches (*híro*) and kebab (*souvlákia*) either to take away or eat at a few tables. Estiatória are usually more up-market places than the ubiquitous tavernas.

garlic sauce), *mavromátika* (black-eyed beans), *saganáki* (fried cheese), and *tirópitta* or *spanachópitta* (small pasties of cheese or spinach respectively).

Vegetarians should check that mince meat has not been added to the rice when ordering *yemístes* (stuffed vegetables) or *dolmádhes* (stuffed vine leaves), and even vegetable soups will be made with chicken stock. With the influx of young foreigners the idea of not eating meat through choice is accepted though totally alien: the Greek word for vegetarian is *hortofágos* (grass eater) – which says it all. Fish tends to be expensive everywhere in Greece because catches are smaller due to over-fishing. Worth trying are: *barboúnia* (grilled red mullet), *marídhes* (fried whitebait) and *xifía* (swordfish marinated in oil and lemon and grilled). *Kalamarákia* (deep-fried squid) are excellent when freshly caught, but most restaurant fare comes frozen from the Far East. *Astakós* (lobster) is particularly good in the Ionian Islands: *garídhes* (shrimps or prawns) are always expensive. Meat dishes include *keftédhes* (meatballs), *biftékia* (a sort of rissole-cum-hamburger), *souvlákia* (grilled kebabs), *arní psitó* (roast lamb) and *katsíki* (roast kid). Local Greek fruit is excellent in season with oranges, lemons, apples, pears, apricots, peaches, nectarines, grapes, figs, melons and pomegranates fresh off the tree.

Greeks have a very sweet tooth and buy cakes and pastries from a **zacharoplastía** (patisserie-cum-café) rather than at a restaurant. Desserts include *baklavá* (a nut-filled phylo pastry soaked in sugar syrup) and *kataífi* (chopped walnuts and honey wrapped in shredded wheat). The **galaktopolía** sell dairy products including *kréma* (custard), *rizógalo* (rice pudding) and locally made *yiaoúrti* (yoghurt – the best is made from ewe's milk). Try the zacharoplastía and galaktopolía for a far better continental breakfast than the regulation fare in hotels. Honey, usually of the very runny kind, is a great favourite and often served poured over creamy, fresh yoghurt. Useful lunchtime or picnic stand-by's are the larger versions of the pasties mentioned above made with phylo pastry: *tirópitta*, filled with minted cheese, *spanachópitta* with spinach and cheese, or *eliópitta* with black olives. *Féta* is the best known Greek cheese – if buying loose, ask to try a bit since some can be very salty; *graviéra* is a locally made Gruyère-type cheese.

> ### RETSINA
>
> Retsina is said to be an acquired taste (easily done with a little determination and long drawn-out island meals). Produced for around 3000 years, it probably originated when wine jars were sealed with plaster and pine resin – the latter was claimed to 'preserve' the wine. Detractors describe it as like drinking turpentine (they imagine) … Most retsina today comes from Attica.

Drinks

Many foreign **beers** are made in Greece under licence; they are cheaper by the bottle than canned. **Oúzo**, the aniseed-based spirit, is synonymous with Greek holidays. It is served with iced water which is tipped into the oúzo turning it milky-white – the smoothest brands come from Samos and Lesbos. Greek **brandies** – the best known is *Metaxás* – are heavier and rougher than their French counterparts but very good as the base of drinks such as brandy sour or Alexander. Greece was famed for its **wine** in antiquity but whereas Bulgaria and Hungary have been quick to use Australian expertise to revolutionize their wines, the same is only just beginning to happen in Greece. Nevertheless, Greek wines drunk on their home ground are perfectly acceptable although you may find that they do not 'travel' and are better as a holiday memory. Every tourist resort has shops selling sundry locally produced 'liqueurs'.

Below: *Oúzo is traditionally served with mezédhes. Genuine oúzeries are hard to find but worth the effort.*

2
Athens and Attica

With thousands of direct flights direct from the UK and mainland Europe to all the most popular islands, there is no real need to use Greece's capital city as a gateway to the islands, and only a relatively small number of visitors do so.

That said, **Athens** has much to recommend it: world-class museums and archaeological sites, great streetlife, fine hotels and restaurants and clear seas and sandy beaches within minutes of the city centre.

A fine new airport, an enviable new metro and tram system, and new pedestrian areas in the historic centre are the legacy of a modernization drive leading up to the 2004 Olympic Games. Newer buses and tighter pollution controls have also curbed the summer smog.

Greek **tourism** started here more than 2000 years ago, when curious Romans visited the city to see its legendary temples; in the 19th century, it became one of the world's very first package tour destinations.

But Athens has lost out to the islands in the tourism numbers game – of around three million Britons who visit Greece each year, for example, fewer than 200,000 visit the capital, and with plenty of flights direct from most European cities to the most popular islands it is no longer so important as a gateway to the Aegean.

More than four million people (almost half the country's total population) live in the greater Athens area, which has spread across most of the northern half of the Attica peninsula and extends for isles along its western coast.

DON'T MISS

★★★ Acropolis: surmounted by the Parthenon – the world's best known temple.
★★★ Temple of Athena Nike: the smaller but elegant challenger to the Parthenon on the Acropolis.
★★★ Plaka: the lively area below the Acropolis filled with a variety of restaurants, cafés and shops.
★★ National Archaeological Museum: a veritable treasure house of Greek history.
★ Erechtheion: with statues of the Caryatids.

Opposite: *The Parthenon symbolizes the glory of Classical Greece; it makes a memorable spectacle.*

Athenians call their city 'the big village' – reflecting the fact that many are first- or second-generation migrants to the city from the islands or the mainland mountains and maintain close family links with their ancestral isle.

For visitors, however, Athens means the historic city centre around the original core of the city, the hill of the **Acropolis**, first settled more than 5500 years ago.

The **Plaka** district, beneath the Acropolis, is a jumble of narrow streets, ancient ruins, cafés and restaurants, overlooked by the white columns of the **Parthenon** and the **Temple of Athena Nike**.

For those who want to combine city sightseeing with swimming and sunbathing, a string of resort suburbs and luxury hotels stretches south from **Glifada**, only 20 minutes from the city centre, to **Vouliagmeni** and the so-called 'Apollo Coast' of the Attica peninsula.

CITY SIGHTSEEING

Platia Syntagma (Constitution Square) is the heart of the modern city. The square was recently given a face-lift, with fountains, shade trees and a new metro station in time for the 2004 Olympics. It is overlooked by the imposing 19th-century façade of the **Vouli** (Parliament Building), where elite soldiers of the evzone regiment guard the **Tomb of the Unknown Soldier**. Between the square and the prominent crag of the Acropolis with its ancient temples is the picturesque **Plaka** quarter – a maze of small streets crammed with shops, souvenir stores, cafés and restaurants. Between **Syntagma** and **Monastiraki** is the 'new' flea-market area, several blocks of up-market tourist shops selling designer clothes, jewellery and leather goods, while west of Monastiraki is the 'old' flea-market area, a labyrinth of junk shops and souvenir stalls. North and west of Monastiraki, the former **Psiri** district has become the city's trendiest nightlife area.

THE ACROPOLIS

Rising 100m (328ft) from the **Plain of Attica**, the steep-sided Acropolis is a natural fortress which attracted the earliest people to settle the site of modern Athens. After the sack of Athens by the Persians in the early fifth century BC (and their subsequent decisive defeat at Salamis, within sight of the Acropolis), the charismatic leader **Pericles** ushered in a golden age for the city by commissioning magnificent new temples dedicated to the goddess **Athena**, protectress of the city.

The **Acropolis Museum** closed in July 2007 when all its exhibits were moved to the **New Acropolis Museum**, which is scheduled to open at the end of 2008. Acropolis site and temples open daily 08:00–19:00.

The Parthenon ★★★

Greece's most evocative symbol, the Parthenon with its 46 Doric columns of white marble was the work of the sculptor **Phidias**, between 447 and 432BC. It housed an 11m (36ft) ivory statue of **Athena Parthenos**, and its columns and foundations are designed to create an impression of great size. A frieze of carvings of gods, heroes and epic battles surmounted the columns. The Parthenon remained almost intact until 1687, when it was severely damaged by the explosion of a Turkish powder store during the Venetian siege of the city.

Below: *Bustling Plaka, a maze of streets, small shops, cafés and restaurants in what remains of Medieval Athens.*

Temple of Athena Nike ★★★

Like the Parthenon, this temple – built by **Kallikrates** around 478BC – was dedicated to Athena and housed a colossal statue of her in her guise as **Nike**, the goddess of victory. The statue was wingless, so that Victory could never leave the city.

The Erechtheion ★★

Built around 421BC, the Erechtheion sheltered a sacred olive tree planted by Athena. Six statues of maidens (caryatids) supported the roof. One was removed, along with the Parthenon friezes, by Elgin, and is now in the British Museum. The others are in the Acropolis Museum, along with a range of other artefacts from the Acropolis site.

AROUND THE ACROPOLIS

Two ancient theatres, the fourth-century-BC **Theatre of Dionysos** (open 08:30–15:00 Oct–Mar; 08:00–19:00 Apr–Sep) and the **Theatre of Herodes Atticus** (same hours as Acropolis) – named after the teacher and philanthropist who financed it in 161AD – are built into the southern slope of the Acropolis.

Below: *The Odeon of Herodes Atticus (AD161) is the focus of the annual Festival of Athens and a venue for theatre, ballet and classical music.*

North of these, lies the **Ancient Agora** or marketplace (open daily 12:00–19:00 Mon; 08:00–19:00 Tue–Sun; daily 08:00–15:00 in winter) where high points include the **Theseion**, a Doric temple dedicated to Hephaistos – god of smiths and armourers – and the

Agora Museum, (it is open daily 08:30–14:45) which is housed in the magnificently restored second-century-BC **Stoa of Attalos**.

The pillars of a great stone portico and a jumble of tumbled columns marks the site of the **Roman Agora**, where the most prominent landmark is the octagonal **Tower of the Winds**, built in Roman times as a sundial, water clock and weather vane.

About 1km northwest of the Ancient Agora is the **Kerameikos** (ancient cemetery) (open 08:30–15:00 Tue–Sun) with remarkable tombs from the fourth and fifth centuries BC and also an interesting museum housing finds from Greek and Roman graves.

East of the Acropolis, the **Monument of Lysikratous** on **Platia Lysikratis** commemorates the patron of a trophy-winning 4th-century choir and across busy Leoforos Amalias 15 huge pillars mark the **Temple of Olympian Zeus** (open 08:30–15:00 Tue–Sun) and **Hadrian's Arch** marks the boundary of the new Roman town built by the Emperor Hadrian.

MUSEUMS

The **National Archaeological Museum** has received an extensive renovation and its superb collection of finds from the ancient Hellenic world is better displayed than ever before. Among its treasures are the golden 'Mask of Agamemnon' from Mycenae, Cycladic sculptures, and the bronze statue of Poseidon discovered on the sea-bed off Evia. (Tossitsas 1, open Tue–Fri 08:00–19:00, Mon 12:30–19:00.)

Three interesting smaller museums stand on Vasilias Sofias Avenue, just north of the Parliament building. The recently renovated **Benaki Museum** (Vas. Sofias 22, open Mon, Wed, Fri and Sat 09:00–17:00) houses Emmanuel Benaki's personal collection of art and archaeological finds from both the Byzantine and the Ottoman eras.

The nearby **Museum of Cycladic Art** (open Mon, Wed, Fri and Sat 10:00–16:00) contains remarkable stone figurines from the civilization which flourished in the Cyclades islands from 3200BC onwards; and the **Byzantine Museum** (open 08:30–15:00 Tue–Sun) houses work from early Byzantine churches.

PIRAEUS

The grimy, busy freight and ferry port of Athens is linked to the city centre by metro and bus (from Omonia, Syntagma, Akropoli, Monastiraki and Thissio stations – allow 30

SHOPPING

Try the Sunday morning flea markets held in **Monastiraki**, **Thissio** and **Piraeus**. **Plaka** also has an abundance of shops geared to the tourist trade. Street markets (*laiki agora*) run in various Athenian neighbourhoods on different days of the week and sell a wide range of household goods and foodstuffs.

The main museums (**Benaki**, **Cycladic Art** and **National Archaeological**) have shops with high-quality reproductions and original art on sale. For antiques, there is a collection of shops on **Adhrianou**.

WOLF MOUNTAIN

It is a long time since the four-footed genuine creatures howled on **Lykavittos** – Wolf Mountain. From **Kolonaki Square** – patronized by the wealthy – it is a long climb on foot but a funicular railway takes you up to the **Chapel of St George**. From up here is a superb view over Athens and Piraeus, atmospheric clarity permitting.

Above: *Evia's cafés and restaurants have remained haunts for local people.*

minutes) and has its own main-line rail station. Ferries and catamarans to Crete, the Cyclades and Dodecanese leave from the main quays opposite the metro station; hydrofoils to the Argo-Saronic isles go from Zea Marina (2km east of the station). Boats to more distant islands (such as Samos, Rhodes and Crete) tend to leave early morning or late evening. Piraeus is a dull place to hang about – time your departure so you don't have to.

RAFINA

Rafina, on the east coast of Attica, is only 20 minutes from Athens international airport so is convenient for those who want to go straight to the islands without visiting Athens. It's a much more congenial place to spend a night than Piraeus, with rows of excellent fish restaurants along its harbour-front (Athenians come here just to eat), and a decent if unspectacular sandy beach. Catamarans depart (mostly morning and evening) for the Cyclades, Sporades, Northeastern Aegean isles and Evia. It is a convenient place to stay overnight before flying home.

AROUND ATTICA
Vravrona

About 8km (5 miles) south of Rafina, the ancient temple of **Artemis** at Vravrona, where a cult of virgins guarded the shrine, is also worth seeing. The temple dates from the fifth century BC, but finds in the small museum are from as early as 1700BC. Temporarily closed.

Marathonas

An earthen mound rising from the seaside plain of Marathonas, 20km (13 miles) north of Rafina, marks the common grave of the Athenian warriors who fell in battle against the invading Persians in 490BC – a victory for the Greeks against huge odds. A museum displays finds from the battlefield and other sites and there's a good, 2km (1.5-mile) sandy beach (where the Persian fleet led by Mardonios landed). Museum temporarily closed; site open Tue–Sun 08:30–15:00.

A HAVEN

A taxi or bus (no. 224) from Akadhimias leading from the northwest corner of Syntagma can take you to the suburb of Kaisariani and a half-hour walk to the **Byzantine monastery** at the foot of **Mt Hymettus**. Just beyond the monastery, paths lead under pine trees on to the slopes of Hymettus, which in spring becomes a haven for over 20 species of wild orchid, a bewildering array of other plants, tortoises and singing birds. On a clear day a tiny, white Parthenon is visible from here, far below on the distant Acropolis.

Ramnous

Before sailing or flying home, visit the peaceful ruins of ancient Ramnous, about 5km (3 miles) north of Marathonas, where the marble columns of a classical temple lie toppled among the trees. Open daily 12:00–19:00 Mon, 08:00–19:00 Tue–Sun (daily 08:00–14:30 in winter).

Evia

Evia is Greece's second largest island after Crete, but is so close to the mainland for most of its 175km (109-mile) length that it hardly feels like an island at all. A swing bridge connects it with the mainland at **Halkida**, about 80km (45 miles) north of Athens, and there are ferries to its southern ports, **Marmari** and **Karystos**, from Rafina. Southern Evia is dominated by 1398m (4587ft) **Mt Ochi** and central Rafina by 1943m (6375ft) **Mt Dirfis**.

Halkida is an unappealing industrial port, with an **Archaeological Museum** (open 08:00–14:30 Tue–Sun) and a **Byzantine Museum**, housed in a former 16th-century mosque. The rapidly changing currents of the **Evripos Channel** between Halkida and the mainland are a local curiosity. **Edipsos** and **Loutra Edipsou**, close to the island's northwest tip, have decent beaches and numerous hot springs. **Rovies**, 20km (13 miles) south of Edipsos on the west coast, has a scattering of beach hotels. Opposite Rafina, **Karystos** is overlooked by the ruins of **Kastello Rosso**, a Byzantine fortress built in 1030. Evia's south and east coasts are almost undeveloped, with some fine, empty beaches, a scattering of small villages and very few places to stay. Midway along the east coast, **Kimi** is the largest village, with ferries to **Skyros**, southernmost of the Sporades island group.

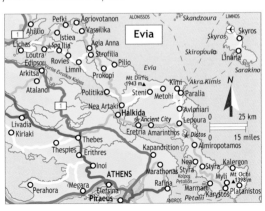

Mt Dirfis

The lower slopes of Mt Dirfis (1745m; 5725ft), the highest peak in Evia, are clothed with cool woods of chestnut and pine. A road crosses the shoulder of the mountain to **Stropones** providing the easiest access to the summit ridge. Above, crocus, white peonies and scarlet lilies are three of the botanical treasures on view.

From **Steni**, with its waterfalls and wooden mountain houses, there is a marked path and challenging walk to the summit.

Athens and Attica at a Glance

BEST TIMES TO VISIT

In **Spring** (Mar–May) the light shows the ruins at their best, wildflowers bloom and the weather is warm. In **Summer** (Jun–Sep) Athens can be dusty and smoggy during the day – evenings are cooler and the Athens Festival hosts various music concerts in the city. **Winter** (Nov–Jan) can be cold but October is warm and clear.

GETTING THERE

Direct **flights** daily to Athens from London, Manchester, Glasgow, New York, Sydney and all European capitals. All flights arrive at Elevtherios Venizelos International Airport. Olympic Airlines and Aegean Airways connect Athens with islands including Chios, Corfu, Crete, Ikaria, Karpathos, Kos, Kefalonia, Kythera, Lesbos, Limnos, Naxos, Rhodes, Samos, and Skiathos. Other domestic airlines incuding Aegean Airlines also link the capital with the islands. **Rail** travellers can reach Athens via Croatia, Serbia and the former Yugoslav republic of Macedonia (tickets and rail passes in the UK, Canada and USA from www.raileurope.com) or via Hungary, Romania and Bulgaria, but most prefer to go via Italy, with ferries operating from Ancona, Bari, Brindisi, and Venice to Igoumenitsa and Patras on the west coast, as well as to Corfu, and coach connections to Athens (also a port of call for luxury cruise ships).

GETTING AROUND

Airport express **buses** E94, E95 and E96 connect the airport with Ethniki Amyna metro station, Syntagma Square in the city centre and Piraeus. Athens is well served by buses and has an air-conditioned metro system, plus overground tram system. Metro Line 1 connects Piraeus with the city centre and northern suburbs; Line 2 runs from Dafni and connects Acropolis, Syntagma and Omonia with the mainline Larissa rail station. Line 3 runs from Monastiraki on the edge of the Plaka, connecting with airport buses at Ethniki Amyna. Tickets for buses, trams and metro can be bought at stations or from períptero kiosks in the city.

WHERE TO STAY

Athens
LUXURY
Andromeda Athens, Timolonteos Vassou 22, tel: 21064 15000, fax: 21064 66361. Elegant boutique hotel in the heart of the embassy district.
Grande Bretagne, Vasileos Georgiou 1, tel: 21033 30000, fax: 21032 28034. Luxurious rooms, prestigious restaurant and a rooftop pool terrace.
Kefalari Suites, Pendeli/Kolokotroni 1, tel: 21062 33333, fax: 21062 33330. Colourful, quirky all-suite hotel; posh suburb; northern Athens.
Life Gallery, Thiseos 103, tel: 21062 60400, fax: 21062 29353. Ultimately luxurious, stunningly designed new hotel

in the affluent Ekali area.
The Margi, Litous 11, Vouliagmeni, tel: 21089 29000, fax: 21089 29143. Superb resort hotel in a seaside suburb.
Sofitel Athens Airport, Elevtherios Venizelos International Airport, tel: 21035 44000, fax: 21035 44444. Modern business hotel; excellent facilities.
MID-RANGE
Acropolis Select, Falirou 37-39, tel: 21092 11610, fax: 21092 16938. Stylish and comfortable – ask for a room with a view of Acropolis.
Dorian Inn, Peiraios 15-17, tel: 21052 39782, fax: 21052 26196. Central, comfortable large hotel, recently renovated.
Hotel Achilleas, Lekkas 21, tel: 21032 16777, fax: 21032 16779. Conveniently located.
BUDGET
Acropolis View, Webster 10, tel: 21032 25891, fax: 21032 50359. Comfortable, friendly hotel; overlooks Acropolis.
Candili, Prokopi, Evvoia, tel: 69740 62100, www.candili.co.uk Grand estate near the sea offering luxury accommodation, countryside activities and art and craft courses.
Athens Youth Hostel, Victoros Hugo 16, tel: 21052 34170, fax: 21052 34015. The only official International Youth Hostel Association property.
Piraeus
MID-RANGE
Savoy Hotel, Iroon Polytechniou 93, tel: 21042 84580, fax: 21042 84588. Comfortable and convenient.

Athens and Attica at a Glance

Rafina
BUDGET
Hotel Corali, Platia Plastiras, Rafina, tel: 22940 22477, fax: 21069 23457. Convenient – a short walk from the ferry port.

Evia
LUXURY
Thermae Sylla Hotel, Edipsos, tel; 22260 60100, fax: 22260 22055, www.thermaesylla.gr Greece's best spa hotel in a stunningly restored 19th-century building.
MID-RANGE
Elaionas, Rovies, tel/fax: 22270 71619, www.eleonashotel. com Five cottages and apartments with individual balconies or courtyards set among olive trees.
BUDGET
Alexandridis, Rovies, tel: 22270 71272, fax: 22270 71226, www.anzwers.org/ trade/alexandridis Low-rise hotel with a pretty garden, built around a quiet courtyard.

Plaka
Platanos, 4 Diogenis, tel: 21082 20666. Unpretentious but very good food.
Byzantino, 18 Kidathineon, tel: 21032 27368. Tables under trees; excellent value.
Eden, 12 Lissiou/Mnissikleous, tel: 21032 48858. Good vegetarian moussakas and quiches.
Around Plaka
Daphne's, 4 Lysikratous, tel: 21032 27971. Greek and international cuisine.
Milton's, Adrianou 91, tel:

21032 49129. Stylish, modern Mediterranean restaurant – Greek food with a twist.
Tou Damigou Bakaliarakia, Kidathinon 41, tel: 21032 25084. Traditional Plaka basement restaurant; specializes in cod and chips Athenian style.
Exarchia and Koukaki (near the Archaeological Museum).
Embraces the area where food is good and reasonably priced.
To Meltemi, 26 Zinni, Koukaki. A genuine oúzeri, excellent seafood *mezédhes*.
I Gardenia, 29 Zinni, Koukaki. Cheap, traditional casserole dishes. Wine from the barrel.
Rafina
Ioakeim, on the harbour, tel: 22940 23421. Great seafood.
Sta Kala Kathoumena, Platia Plastiras, tel: 22940 25688. Elegant taverna serving dishes from all over Greece.
Evia
Stavedo, Karaoli 1, Halkida waterfront, tel; 22210 77977. Elegant and sophisticated restaurant; good wine list.
Tsaf, Platia Favierou, Halkida, tel: 22210 80070. Traditional oúzeri; excellent seafood.
O Glaros, Agios Nikolaos, Edipsos, tel: 22260 60240. Fish restaurant on waterfront 2km

(1.6 miles) from town centre.

Around Syntagma Square and Plaka there are many travel agents offering trips to the major Greek sites including **Delphi**, **Corinth** and further afield into the Peloponnese for **Epidavros**, **Mycenae** and **Olympia**. One of the easiest trips is to **Sounion**, south of Athens, and the magnificent **Temple of Poseidon** – either by regular service coach from KTEL terminal (Areos Square) or organized trips (travel agents).

Tourist Police 171.
Piraeus Port Authority, tel: 21042 50229 or 21045 93000.
Rafina Port Authority, tel: 22940 22300.
Elevtherios Venizelos International Airport, tel: 21035 30000.
National Tourist Organisation of Greece, 26A Amalias, tel: 21033 10392, www.gnto.gr
NTOG Piraeus Office, Zea Marina, tel: 21041 35716.
NTO Airport Office, tel: 21035 30445.
Rafina Tourist Information Office, tel: 22940 67622.

ATHENS	J	F	M	A	M	J	J	A	S	O	N	D
AVERAGE TEMP. °F	48	49	54	60	68	76	82	82	76	66	58	52
AVERAGE TEMP. °C	11	11	12	16	20	25	28	28	25	19	15	12
HOURS OF SUN DAILY	5	5	7	9	11	15	12	11	9	6	5	4
RAINFALL ins.	2.12	1.81	1.26	0.8	0.74	0.47	0.16	0.32	0.63	1.73	2.48	2.84
RAINFALL mm	54	46	32	21	19	12	4	8	16	44	63	72
DAYS OF RAINFALL	11	11	10	9	8	5	3	4	4	8	11	11

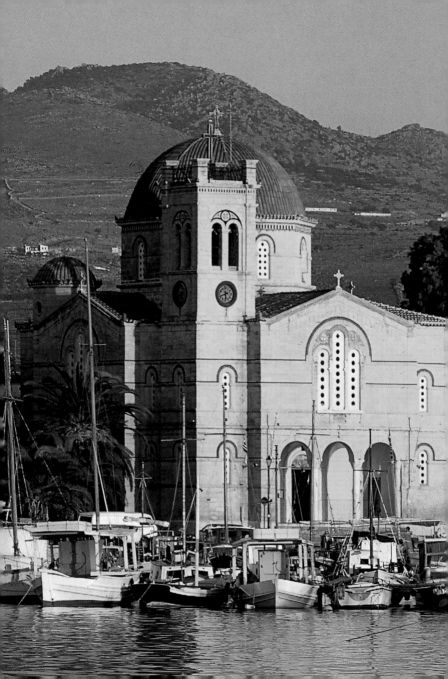

3
The Argo-Saronic Islands

Scattered across the blue waters between Attica and the Argolid peninsula on the southern mainland, the Argo-Saronic isles are a welcome escape from the busy streets of the capital.

Many better-off Athenians have holiday homes on **Aegina**, **Poros**, **Spetses** and **Hydra**, and all four are ideal for a holiday combining the charm of the islands with the excitement of the city. High-speed hydrofoils and catamarans connect the islands with **Piraeus** and **Zea** in 60–90 minutes, making it possible to visit the capital's major sights on a day trip, while Poros lies only 400m (430yd) from the coast of the Argolid, with a shuttle ferry making it an ideal base for exploring the region's great Classical and Mycenaean sites. And as well as some of the most charming of island harbour towns, sandy beaches and pebbly coves, the Argo-Saronic islands also have a handful of archaeological sites, important Byzantine churches and quirky small museums of their own.

These islands have played an important part in Greek history through the ages. Off **Salamina**, the Athenian fleet wiped the Persian Empire from Greek seas in 480 BC, ending the Persian threat to the Hellenic world. Spetses and Hydra were the home ports of powerful trading fleets and lent their armed brigantines to the national cause during the War of Independence, and Hydra was briefly the capital of the new Republic.

While package tourism has diverted the flow of tourism to larger islands with international airports,

ALBANIA
TURKEY
●ATHENS
MEDITERRANEAN SEA

DON'T MISS

★★★ Aegina: the Doric Temple of Aphaia – the best preserved temple on any of the islands.
★★★ Spetses: the town with pavements of black and white pebble mosaics.
★★ Aegina: Perdika with its wildlife sanctuary and fish restaurants; a great favourite with Athenians.
★★ Hydra: Profitis Ilias monastery and the convent of Agios Efpraxia.
★ Poros: Kalavria for the many sandy coves backed by tall pine trees.

Opposite: *The twin domes of the church of Agios Nikolaos is a landmark at Aegina town's waterfront.*

CLIMATE

The weather of the Saronic gulf is the weather of Athens – hot and dry in summer and cold in winter – but without the smog (*nefos*). Summer visitors to Athens and residents use the Saronic islands as an escape to cleaner air.

Aegina, Hydra, Poros and Spetses have retained their authentically Greek flavour, with some fine restaurants and some excellent small hotels. **Salamina** – the largest, closest to Athens and the most built-up of the group – is, however, the ugly sister. Virtually an industrial suburb of the capital, its polluted shores and uninspiring views of the oil refineries of **Elevsina** and shipbreaking yards of **Skaramanga** have no appeal for the visitor.

AEGINA ★★★

The closest of the Argo-Saronic islands to the capital, Aegina is understandably popular with Athenian week-enders as well as with package holiday-makers from the UK and northern Europe. Elegant neoclassical buildings dating from the 19th century line the curving waterfront of its main (and only) town, which is also called **Aegina**.

With an area of only 84km² (32 sq miles), Aegina is an easy island to explore, and has the richest ancient heritage of any of the Argo-Saronic isles. The island's best beaches are at **Aiginitissa**, **Marathonas** and **Perdika**, each of which has a small array of hotels, guesthouses and restaurants.

At the site now known as **Kolona**, a single column marks the site of a 5th-century Temple of Apollo and of the ancient city, once a rival to the power of Athens.

Below: *The Temple of Aphaia on Aegina is the best preserved of all island temples.*

With 25 of its original columns still standing, the superb Doric **Temple of Aphaia**, above the village of **Mesagros**, is the most impressive ancient temple, though now lacking the marble sculptures depicting scenes from the Trojan War which once adorned its pediment and are now in the Pinakothek in Munich. The goddess Aphaia was worshipped almost exclusively on Aegina and was later identified with Artemis.

Many finds from the temple can now be seen in the **Archaeological Museum** in Aegina town.

Only a monumental staircase and two terraces remain to indicate the site of the **Temple of Hellanion Zeus**, on the slopes of **Oros** (534m/1706ft), but it is worth the one-hour hike to the summit for the views over the Saronic Gulf, its islands, and the mountains of the Peloponnese. The monastery of **Agios Nektarios**, in the centre of the island, attracts Orthodox pilgrims. More interesting is the ghost town of **Paleochora**, once the most important settlement on the island. Sacked by corsairs in 1538 and 1654, it was finally abandoned when more peaceful times made it safe to live on the coast. The ramparts of a Venetian fort loom over the old town, and some of its 20 surviving churches – including the **Episkopi** (Cathedral), the **Chapel of the Taxiarchis** (St Michael), and the **Basilica of Agios Anargyroi**.

Offshore, the small island of **Moni** has been established as a santuary for a small flock of Cretan wild goats (kri-kri – *see* p. 10), and the pioneering wildlife sanctuary at **Perdika** also rehabilitates injured wild birds and animals from all over Greece.

AEGINA PEANUTS

Aegina is the main producer of pistachios – locally known as 'Aegina Peanuts'. Canvas sheets are spread beneath the trees in late August and the nuts knocked on to them. After hulling, the nuts are soaked in sea water and dried on flat roofs and terraces in the sun. Local abundance is not reflected in the prices charged in town.

Above: *Elegant stone mansions overlook the harbour on Hydra. Once the homes of sea captains, they have been taken over by artists and other creative folk.*

ANGISTRI ★

Tiny Angistri is only 17km² (7 sq miles) in area and in summer and at weekends its only beach resort, **Skala**, can be very crowded with Athenians as well as package holiday-makers – the latter, ironically, hoping to escape the crowds. Many of the island's 700 inhabitants (like those of Hydra) are descended from Greek-Albanian refugees from Epirus who fled the conquering Turks in the 15th century.

POROS ★★

Most of Poros's 4500 people live in its eponymous main town, which rises in tiers of narrow streets above a pretty natural harbour, looking across the narrow 370m (403yd) strait to Galatas, a small mainland port which is surrounded by citrus groves. Poros is virtually two islands in one, with its main town standing on a near circular headland, **Sferia**, which is connected to the main part of the island, **Kalavria**, by a narrow isthmus. Poros is hilly and wooded, and its best beaches, on the north coast, are linked to the harbour by frequent shuttle-boats. The **Monastery of Zoodochos Pigi** (Lifegiving Spring), in the centre of the island, is the only site worth visiting. Only a few scattered stones remain of the **Temple of Poseidon**, in the northeast.

More exciting sightseeing lies on the mainland, within easy reach of Poros and Galatas. Closest (40km/25 miles northwest of Galatas) is **Epidavros**, the most impressive of Greece's ancient theatres. Seating an audience of 14,000, its acoustics are superb and are brought to life during the annual **Epidavros Festival** (Jun–Aug). Approximately 32km (20 miles) further west are the 3700-year-old tombs and walls of Mycenae, Agamemnon's capital, and the Cyclopean ramparts of Tiryns, dating from the 13th century BC and built of huge stone blocks.

HYDRA ★★★

Tiers of dignified grey stone mansions, rising above a perfect natural harbour, are the legacy of Hydra's golden age as the home of one of the Aegean's most powerful trading fleets – though Hydra's sea captains as happily turned their hands to piracy when offered the chance. Hydra is rocky and barren, and the Greek-Albanian refugees who settled here in the 15th century had no choice but to turn to the sea.

Today, the island is car-free and its coasts and coves can be explored by boat, while inland, donkeys will carry your luggage to your hotel and mules will carry you to the **Convent of Agia Efpraxia** and the **Monastery of Profitis Ilias**, high above the harbour. There is good swimming at **Vlichos** and at **Limioniza** on the south coast, and the ruins of a Venetian fort can be seen above the sea at **Kastello**. Caiques take about an hour to reach **Dokos**, an island of hard marble (marmaropita) where pioneering diver Jacques Cousteau discovered the site of a 300 year old shipwreck.

SPETSES ★

Located the furthest from Athens of all the Argo-Saronic islands, Spetses is long on low-key charm but short on sightseeing and outstanding beaches.

Low-lying and with an area of 22.5km² (9 sq miles), Spetses has an attractive island capital, usually known as **Dapia**, after the central harbour square which is attractively cobbled with black and white pebbles. **Paleo Limani** (the old harbour) is crowded with fishing boats and yachts.

Above: *Spetses is justifiably popular with families for its safe beaches of pebbles and shingle.*

LASKARINA BOUBOULINA

On 2 April 1821, Spetses became the first Greek island to raise the flag of Independence. The island's fleet figured large in battles against the Ottomans under the command of **Laskarina Bouboulina**, the island's lady admiral and mother of six who, if the battle looked more bloody on shore, would abandon ship and set off with her sabre.

She was a skilled tactician and inspirational leader who could outdo any man in the drinking stakes, so it is perhaps to preserve fragile male egos that she is reputed to have been extremely ugly.

The Argo-Saronic Islands at a Glance

BEST TIMES TO VISIT

The islands are particularly pretty in **spring** (Mar to Apr) with a profusion of wild flowers, and this is also a good time for sightseeing in Athens and the mainland, though the water is chilly at this time of year. Throughout the **summer**, and especially over weekends and during Greek public holidays, all the Argo-Saronic Islands (and especially tiny Angistri) can be very crowded. **Autumn** is probably the most perfect time to come here, with fewer visitors and pleasantly warm seas. **Winters** are generally cool and often wet, and virtually all the hotels and other tourist establishments close from Nov until Feb or Mar.

GETTING THERE

Ferries and hydrofoils bound for Aegina, Angistri, Poros, Hydra and Spetses leave from the main Piraeus harbour and from Zea Marina several times daily. Some of the summer services continue from Spetses to mainland ports including Leonidion, Kiparissi, Geraki, and Monemvasia, and also to Kythera island.
Very frequent **shuttle ferries** also link Angistri with Aegina, Poros with Galatas and Methana on the mainland, and Spetses with Ermioni and Portoheli on the mainland.
Ferry lines include:

Angistri-Piraeus Lines, tel: 22970 91433.
Hellenic Seaways, tel: 22980 54007, www.hellenicseaways.gr
Saronic Dolphins, tel: 21042 24980, www.saronicdolphins.gr

GETTING AROUND

On **Aegina** a good bus service runs from town (Platia Ethneyersias) to most villages and sites; caiques link Aegina town with Angistri and Perdika with Moni.
On **Hydra** water taxis operate from the quay to beaches and the islet of Dokos.
On **Poros** there is one bus only, which goes to the monastery and back.
Salamina has a bus service operating between villages.
Spetses relies on horse-drawn carriages, scooters and bicycles (see p. 124 for on-line timetables and bookings).
To explore the **mainland** from Poros, you can rent a car from one of the agencies in Galatas. The usual warnings apply when renting a car: make sure you have full insurance coverage and collision damage waiver and always drive defensively.

WHERE TO STAY

During the high season, rooms are in short supply on Aegina, Hydra and Spetses. Be sure to book ahead – hotels at all price levels can be found and booked on the Greek Travel Pages, www.gtp.gr
Outstanding hotels in this region include:

Aegina
MID-RANGE
Aiginitiko Archontiko, Ag Nikolaou/Thomaidou 1, tel: 22970 24968, fax: 22970 26716. This lovely 18th-century mansion has 12 pretty rooms round two inner courtyards.
Moonday Bay, tel: 22970 61215, fax: 22970 61147. This is a larger resort hotel of around 80 spacious rooms in eight two-storey blocks among pine trees.

BUDGET
Petrino Spiti, tel: 22970 23837. Tasteful suites and studio apartments in an old townhouse in Aegina town, not far from the harbour.

Hydra
Luxury
Bratsera, Tombazi, tel: 22980 53971, fax: 22980 53626. This is the best hotel on Hydra, with a fine restaurant and the distinction of having the only pool in town.
Orloff, Rafalia 9, tel: 22980 52564, fax: 22980 53532. This charming 18th-century mansion was renovated in 1988. It boasts a flower-filled courtyard and has charming bedrooms that are furnished with antiques.

The Argo-Saronic Islands at a Glance

MID-RANGE
Leto Hotel, tel: 22980 53385, fax: 22980 53806. This hotel blends modern facilities with Hydriot style. It offers various interesting arts courses during the summer months.
Miranda, tel: 22980 52230, fax: 22980 53510. This is a charming traditional mansion with its own art gallery. It also has a lovely courtyard garden.

BUDGET
Hydra, Voulgari 8, tel: 22980 52102, fax: 22980 53805. This 13-room hotel has wonderful views overlooking the harbour and offers excellent value for your money.

Poros
MID-RANGE/BUDGET
Sto Roloi,
Hatzopoulou/Karras, tel: 22980 25808. Two apartments and one double room are available in this blue-and-pink 19th-century house. It is centrally located in the heart of the old quarter.

Spetses
Luxury
Zoe's Club, tel: 22980 74447, fax: 22980 72841. Luxury apartments are situated around a large pool on the outskirts of Dapia. This establishment is excellent for families.

All the Argo-Saronic Islands have the usual array of fish restaurants, grills and run of the mill tavernas where you can choose what you want from the kitchen. Some of the more outstanding places to eat include:

Aegina
Agora, tel: 22970 27308. This is Aegina's oldest and best fish restaurant. It stands beside the fish market, so its seafood is guaranteed fresh.
Antonis, tel: 22970 61443. Beside the fishing harbour at Perdika, this famous fish restaurant attracts wealthy Athenians and is one of the few local restaurants that stay open all year round.

Hydra
Porphyra, Hydra Town, tel: 26950 53660. This is one of the more affordable places to eat in pricey Hydra, offering basic Greek cooking.
Kondylenia, tel: 22980 53520. Authentic fish restaurant with tasty local specialities including sea urchins and samphire.
Kseri Elia, tel: 22980 52886.

Simple taverna with excellent meze dishes.

Spetses
Tarsanas, Palio Limani, Spetses town, tel: 22980 74490. Up-market, harbour-side seafood restaurant that attracts droves of Athenian weekenders in summer.
Exedra, tel: 22980 73497. Situated right on the old harbour, this mid-priced fish restaurant has tables beside the sea.

Day trips to Athens by hydrofoil are offered by travel agents on all the Argo-Saronic islands, and all the islands are within easy day-trip reach of each other. Agents also offer one-day cruises combining three or more of the islands. Agencies on Poros also advertise escorted coach tours to Epidavros and Mycenae.

Tourist Police:
Aegina, tel: 22970 27777.
Hydra, tel: 22980 52205.
Poros, tel: 22980 22460.
Spetses, tel: 22980 73744.

SARONIC ISLANDS	J	F	M	A	M	J	J	A	S	O	N	D
AVERAGE TEMP. °F	46	46	52	59	68	75	82	81	75	68	57	50
AVERAGE TEMP. °C	8	8	11	15	20	24	28	27	24	20	14	10
HOURS OF SUN DAILY	5	5	6	7	10	11	12	11	9	7	5	4
RAINFALL ins.	2	2	2	2	1	0.5	0.5	0.5	0.5	2.5	3	2
RAINFALL mm	59	48	50	43	22	12	10	14	14	65	77	55
DAYS OF RAINFALL	11	10	10	9	7	4	2	3	3	8	10	12

4
The Cyclades

Scattered across the central Aegean, the Cyclades are the islands everyone thinks of as quintessentially Greek: tiny, treeless except for olive groves and clumps of tamarisk and oleander, and dotted with dazzling white villages and tiny blue-domed churches. The group takes its name from the Greek word 'kyklos', meaning a circle, as the islands form a loose ring around **Delos**, the sacred island of ancient Greece.

Few islands have more than one main harbour-village, almost always designated on the map with the same name as the island, and almost always known to islanders simply as 'Chora' – 'the village'. Only two dozen of the 50 or more islands are inhabited, and their landscapes are surprisingly varied, from the low, barren hills and sandy beaches of Mykonos to the steeper hills and broader landscapes of **Andros** and **Naxos**, the dramatic caldera of volcanic **Santorini**, the sea-cliffs of **Folegandros** and the weird rock formations of **Milos**.

And the Cyclades appeal to an equally varied audience. **Paros** and **Ios** have been favourite party islands for budget island-hoppers for more than 30 years. **Mykonos** has a thriving gay scene, stylish shops and restaurants and great nightlife. **Santorini** offers magnificent sunsets and some of the best boutique hotels in Greece. **Naxos** and **Andros**, the largest islands in the group, are surprisingly untouched by the package holiday trade, while the rugged landscapes and beautiful traditional villages of **Folegandros** and **Amorgos** attract travellers looking for peace and quiet. Smaller islands, such as **Koufonissi**,

ALBANIA
TURKEY
●ATHENS
MEDITERRANEAN SEA

DON'T MISS

★★★ **Mykonos:** the white-washed town (Chora) and delightful 'Little Venice'.
★★★ **Sifnos:** and its ornate Venetian dovecotes.
★★★ **Santorini (Thira):** its volcanic scenery and Akrotiri with its buried Minoan town.
★★ **Kea:** Ioulis with red pantiled houses and its windmills.
★ **Delos:** island sanctuary of Apollo known for the Delian Lions and many other remarkable antiquities.

Opposite: *The dazzling white buildings and blue domes of Dhia on Santorini typifies the clean lines of Cycladic architecture.*

Above: *Batsi, the main tourist centre on Andros.*

Skinoussa, **Iraklia** and **Anafi**, attract even more determined castaways. **Tinos**, within sight of the fleshpots of Mykonos, is one of Orthodox Greece's most important pilgrimage sites. Closer to Athens, **Kea** is a popular weekend getaway for city-dwellers and many Athenians have second homes here.

The Cyclades are perfect for island-hopping, with few islands more than a couple of hours away from each other and onward connections eastward to the Dodecanese and Northeast Aegean islands and south to Crete.

THE EASTERN CYCLADES

The Eastern Cyclades are the most popular and best known of all the Greek islands stretching from Andros – within sight of Attica – to Santorini, around 100km (62 miles) north of Crete.

Andros ★★

Not far from the mainland, the northernmost of the Cyclades is a long, thin island, rising to a central ridge with a 994m (3260ft) summit, **Apikia**, in the centre of the island, which with an area of 380km² (147 sq miles) is the second largest of the Cyclades. The southern part of the island is steep and rocky, while the north is arid, with spectacularly terraced fields on the mountain slopes.

Though popular with wealthy mainlanders, the island sees relatively few foreign visitors, and its attractions are low-key. **Andros** (Chora), the main town, stands high above its harbour and its streets are lined with the neo-classical mansions of 19th-century shipowners, some of whose descendants are still among Greece's wealthiest shipping dynasties. One of those dynasties, the Goulandris family, has endowed the town with a fine **Archaeological Museum**, famous for the Hermes of Andros (open 08:30–15:00 Tue–Sun) and an enchanting **Museum of Modern Art**, with works by Bouzianis, Manolidis and

Tombros (open 10:00–14:00, 18:00–21:00 Wed–Mon). Several attractive churches – larger and more opulent than the little chapels of less wealthy islands – are also the legacy of Andros's entrepreneurs, and Agios Thalassini and Agios Georgios are well worth seeing.

Most of Andros's tourism centres on **Batsi**, midway along the west coast with mediocre beaches and hydrofoil connections to the mainland. Andros's other ferry port, **Gavrion**, lies 6km (4 miles) north of Batsi and is a somnolent spot with little more than a few tavernas and cheap hotels.

Inland, the ruins of **Paleopolis**, the ancient capital, lie on a hill reached by more than 1000 steps from the modern village of the same name. Other villages worth visiting include **Stenies**, the prettiest village on the island, **Apikia**, and **Menites**, both known for their mineral springs.

Tinos ★

Tinos is separated from Andros by a channel less than 1km (0.6 mile) wide and its main claim to fame is the miraculous icon of **Agios Loukas** (St Luke) in the overblown **Church of Panagia Evangelistria** which dominates the town. Thousands of pilgrims visit the island annually in hope of being cured of all manner of ailments.

Miracles aside, Tinos is a pleasantly quiet island, with sleepy hill villages, hundreds of ornate whitewashed Venetian dovecotes (peristerouias) in mountain valleys, undemanding hill-walking and clean if unspectacular sandy and pebbly beaches at **Agia Fokas** (closest to Chora), **Xeres**, **Porto**, and **Kionia**, all on the southwest coast. For more peace and quiet, head for **Ormos Panormos**, a pebbly bay on the northeast coast,

> ### THE GOLDEN ICON
>
> An icon believed to have been painted by St Luke was discovered on Tinos in 1822. Following a vision Sister Pelagia, a nun at Kechrovouni Convent, was directed to the location of the icon beneath a rock. The icon, known as the **Megalochori** (Great Grace), is now encased in gold and reputed to have great healing powers.

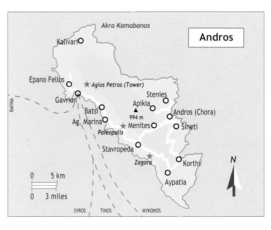

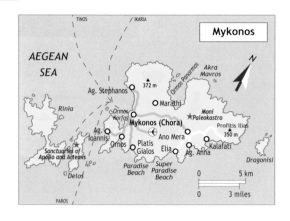

where there are a few simple guesthouses and tavernas. Tinos has few historic relics: above **Kionia** are the scattered remains of a temple to Poseidon and Amphitrite, while on the slopes of **Mt Exoumbourgo** are the frowning ramparts of an impressive Venetian fortress. Built into its walls are blocks and column drums looted by the Venetians from the ruins of an ancient Greek settlement.

Mykonos ★★★

Mykonos was the first of the Cyclades to appear on the tourism map. As early as the 1950s its gorgeous white Chora – one of the most striking in the Aegean – attracted painters and bohemians, and by the 1960s it had become a haven for jet-setters and millionaires. It still attracts its share of the super-rich, judging by the number of private jets landing at its airport, but the best sandy beaches in the Aegean, fine hotels and restaurants and great nightlife attract a wide audience, from young British package tourists to a style-conscious gay crowd and a regular influx of American and European cruise passengers.

Below: *Picturesque old houses stand on the seafront in Mykonos town.*

Chora has few attractions other than boutiques, bars, clubs and restaurants, but its **Archaeological Museum** (open 08:30–15:00 Tue–Sun) has finds from the ancient cemetery on neighbouring Rinia and the Aegean Maritime Museum has an interesting collection of prints, maps and models of sailing ships that were still plying the Aegean in the mid-20th century. A row of dilapidated stone windmills overlooks the harbour, and the painted wooden balconies of the houses which rose straight from the sea below have earned them the nickname 'Little Venice'. Mykonos's fine south coast beaches are crowded in summer, with sun-loungers and parasols cramming every inch of **Ornos** (nearest to Chora, but tiny), **Platis Gialos** (larger, but equally busy) and the two nudist beaches nicknamed '**Paradise**' and '**Super Paradise**' – the latter once a mainly gay beach, now mixed. **Agia Anna** and **Kalafati** beaches, further east, are a little less crowded.

Above: *The stylized Lions of Delos have 'guarded' the shrine to Apollo for more than 2000 years.*

Delos ★

Uninhabited, and off-limits for an overnight stay, the whole of the little island of Delos is designated an archaeological site. This was the most sacred shrine of the sun god Apollo and his twin sister, Artemis and after Delphi, was the holiest spot in the ancient Greek world.

Boats leave frequently from Mykonos, and Delos's antiquities are revealed as soon as you land. By Roman times Delos had become an important trade centre, and visiting merchants left offerings to the Lares Competales – the gods of commerce – at the **Agora of the Competales**, beside the landing stage. The road which led to the three temples which made up the **Sanctuary of Apollo** – the largest of which was built in 476BC – was originally lined with statues, but the **Terrace of the Lions** with its five stone guardians is the most striking sight on Delos. Also to be seen is the ancient theatre, which could seat 5500, the remains of an impressive arched cistern, and the foundations and remnants of the mosaic floors of a row of Hellenistic villas.

Above: *Campaniles (bell towers) are a feature of many island churches.*
Opposite: *Windmills like this one in Antiparos are a familiar sight near ports.*

Syros ★

In the 19th-century **Ermoupoli**, capital of Syros, was the most important seaport in Greece, and it is still a busy commercial entrepot – very different from the quaint little fishing harbours of neighbouring isles.

The legacy of its wealthy heyday can be seen in opulent public buildings including the town hall, which dominates one side of the main square, **Platia Miaoulis** (named after the island's most famous son, a prominent admiral of the Greek fleet in the War of Independence). Behind the square is the **Apollon Theatre**, a copy of La Scala in Milan, and the grand mansions of the **Vaporia Quarter** form an amphitheatre above the sea. Syros was a Venetian stronghold, and the Venetian legacy lingers: to this day, the island is the home to one of Greece's few Roman Catholic communities, and the **Catholic quarter** is a labyrinth of small streets with several remarkable churches, including the Cathedral of Agios Georghios, the Church of Agios Nikolaos, and the Capuchin Convent of Agios Ioannis. Few foreigners disembark at Ermoupoli, and even fewer explore the island's hinterland. For those who do, there is good walking in spring and autumn, and a scattering of small, secluded beaches along the north coast.

Paros ★★

Paros is the first stop for thousands of island-hoppers each year. In the middle of the Cyclades, it is blessed with a pretty, lively main village, **Parikia**, some good beaches, and frequent ferry connections in all directions, making it the ideal place to plan your next step. Paros was famed for its marble, which built many of the temples of ancient Greece and was last used in the making of Napoleon's tomb – the old quarries can be seen at **Marathi**, east of the main village. Parikia's long harbour front is a mass of hotels, restaurants and travel agencies, but a few steps back from the harbour the heart of the village is a charming maze of whitewashed houses and narrow streets. Just west of the harbour is the beautiful 'Church of a Hundred Doors' (Ekatontapyliani), built in the 6th century and rebuilt many times since then. Nearby, the remains of a Frankish tower are studded with blocks and column drums recycled from earlier classical and Hellenistic buildings. The **Archaeology Museum** (it is open 08:30–15:00 Tue–Sat) displays some of the marble tablets on which are carved one of the earliest histories of Greece, the **Parian Chronicles**; the rest are in the Ashmolean Museum in Oxford.

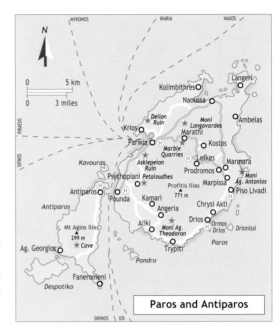

Paros and Antiparos

Many visitors never get beyond Parikia, its bars, clubs, and perfectly adequate town beach. For those who do, Paros offers far better beaches at **Drios** and **Chrysi Akti**, on the east coast of the island, while **Naoussa**, on the north coast, offers a growing number of places to stay and ferry links to neighbouring Naxos.

Antiparos ★★

Antiparos, separated from Paros by a narrow channel and reached by shuttle-ferry or day-excursion boat from Parikia, is very different from its bigger neighbour – and something of a well-kept secret that is bypassed by many visitors to Paros. Its single village has a choice of cheap and cheerful pensions, small hotels, and tavernas and there are long, uncrowded sandy beaches at **Kastro** and **Agios Georgios** on the south coast – though, in the absence of public

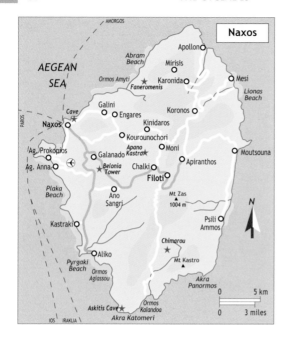

transport, you'll need to rent a scooter to get there. In the centre of the island, a 70m (230ft) deep cave has been attracting visitors since antiquity – it was reputedly visited by Alexander the Great, and more recently by Lord Byron, who indulged his usual fondness for carving his name into its walls.

Naxos ★★

The largest of the Cyclades, at 448km² (173 sq miles), Naxos hardly needs tourism. One of the most fertile islands in the Aegean, its populous hinterland is dotted with prosperous farming villages and produces rich crops of potatoes, citrus and grapes. The island capital was from the 13th to 17th centuries the seat of the Venetian Sanudo dynasty, whose Duchy of Naxos also included Paros and Santorini. Their citadel, **Kastro**, dominates the town and its harbour from a low hilltop, and Venetian coats of arms can be seen on 19 old houses within its crumbling walls, along with 13th-century Catholic cathedral. Its Orthodox counterpart, **Zoodochos Pigi** (Cathedral of the Lifegiving Spring) is in the heart of the **Bourgos quarter**, or lower town. On the slopes of **Mt Zas** (1004m; 3294ft) at the head of the lovely **Tragea Valley**, lies **Filoti** with two attractive churches built mainly from marble. Good walks lead from here to the summit or to the Hellenistic tower of **Chimarou**. Naxos's best beaches lie south of Chora, where **Agia Anna** has developed

Below: *A huge, stone figure (Kouros) of Apollo lies near the entrance to a marble quarry close to the lovely village of Apollon.*

into a small-scale resort (mainly because of its proximity to the island's airport) with a long stretch of sand and pebbles running south to **Kastraki**, **Aliko** and **Ormos Agiassou**.

Koufonissi, Donoussa, Skinoussa and Iraklia ★★★

Between Naxos and Amorgos to the southeast lies a mini-archipelago of tiny islands that apart from a summer influx of devotees see few visitors. Most accommo-

Above: *Chozoviotissa monastery on Amorgos set against a steep cliff face.*

dation is in family-run guesthouses. Relatively difficult to get to and get away from (small and somewhat irregular ferries operate from Naxos and Amorgos) these are islands for castaways, with idyllic beaches, clear water and not much else – but beware, they can become crowded in July and August.

Donoussa, due east of Naxos, has some of the best beaches and is hillier than its siblings, with good walking and sea views. **Iraklia** has a pretty old Chora and a cave noted for its rock formations. The twin **Koufonissi** islets, with a joint area of just 3.8km² (1.5 sq miles) have long beaches at **Finikas** and **Pori**, and unspoilt **Skinoussa** is covered with low hills crowned by three stone windmills.

Amorgos ★★

Barren, wild, unspoiled and friendly, Amorgos (area 134km²; 52 sq miles) appears in a handful of holiday brochures but is largely undeveloped. **Katapola**, the main port, lies tucked into a bay below the hilltop Chora, a dazzlingly pretty and unspoilt white village ringed by crumbling Venetian ramparts that enclose 40 tiny chapels – one of which, Agios Fanourios, is claimed to be the smallest in Greece. The remnants of ancient Minoa, above Chora, are hardly worth the walk, but the giddy path to the **Monastery of Chozoviotissa** (open daily 08:00–14:00 and 17:00–20:00) is a must. Seemingly glued to the cliff face, this whitewashed, fortress-like shrine is one of the most impressive in Greece. From the

monastery, a demanding cobbled mule-path (*kalderimi*) leads over hillsides and through ghost villages to **Egiali**, the island's second port, at the mouth of a dry valley dotted with white villages and overlooked by a row of seven hilltop windmills. The nearby beaches are one of the Aegean's well-kept secrets, and Egiali has a good choice of places to stay and surprisingly lively summer nightlife.

Ios ★★

Ios has been party central since the 1970s. The first waves of bohemian hedonists were drawn by its pretty village and great beaches, but tourism today is better organized. Comfortable camping and affordable guesthouses have replaced sleeping rough on the beach, and although most visitors come in search of a good time, Ios manages to avoid the alcohol-fuelled excess that spoils some resorts on bigger, package holiday islands. **Milopotas**, the island's biggest beach, ranks with the best in the Cyclades, and the island's Chora – on a hilltop above the port of **Gialos** – still has great charm. Over-30s, however, should consider visiting in spring, before the party season really gets under way.

Sikinos ★

Tiny, barren Sikinos is unspoiled and the perfect spot for those whose idea of holiday heaven is a book, a beach, and the prospect of a bottle of wine and a simple supper. Its rocky hills are intensively terraced and covered with tiny vineyards, barley fields and olive groves, and donkeys and mules are still in regular use. **Kastro**, its half-ruined

---

THE BACK ISLANDS

Between Amorgos and Naxos lies a chain of six small islands with idyllic beaches and superb walks. Most accommodation is with island families and it helps to know some basic Greek.

• **Donoussa:** difficult to reach but lovely beaches and walks.
• **Iraklia:** has an old Chora and a cave with impressive stalactites and stalagmites.
• **Keros:** once a centre of Cycladic culture.
• **Koufonissi:** claims to be the smallest inhabited island pair (3.8km²; 1½ sq miles) in Greece – long beaches at Finikas and Pori.
• **Skinoussa:** completely unspoilt with nine low hills and three windmills.

main village, straddles two hills above a somnolent fishing harbour, and its main square is surrounded by balconied, whitewashed houses. Accommodation is limited to a few small pensions by the harbour, where there are also simple cafés and tavernas.

Folegandros ★★
Sheer cliffs rising from the sea give Folegandros (total area 32km²; 12 sq miles) its palpable air of wildness. **Chora**, the capital, is a delightful village perched 300m (984ft) above the sea and reached by a switchback road from the ferry harbour. The 13th-century **Kastro quarter** of Chora, built by Marco Sanudoi, Duke of Naxos, is a maze of alleyways, balconied houses and surprisingly creative and stylish craft shops. Folegandros lacks big beaches, but there is good swimming in crystal water at the tiny bay of Angali, a stiff one-hour hike from Chora, where there are also a couple of simple tavernas.

Santorini (Thira) ★★★
Santorini is, by far, the most immediately striking of all the Greek islands. Its black, red and grey volcanic cliffs leap almost 300m (1000ft) from a deep blue caldera – the product of a volcanic eruption that ripped the island apart some 3750 years ago. This titanic event endowed the island with a rich volcanic soil, and almost every square metre of its hinterland is used to grow plum tomatoes and Assyrtiko vines which produce some of Greece's best wine. Vulcanism also produced the black sands of the east coast beaches, **Kamari** and **Perissa**, which are divided by the ridge of **Profitis Ilias**, crowned by the monastery of the same name and by the scant ruins of **Ancinet Thira**. More fascinating ruins are being unearthed at **Akrotiri**, the island's southern tip, where archaeologists have discovered the remnants of an important Minoan-era settlement buried beneath the volcanic ash. In 2004 the site closed for the installation of a high-tech protective canopy but should reopen by 2006.

ISLAND REMNANTS

Thirasia is a reminder of Thira before the tourist invasion and, like Nea Kamena, Palia Kameni and Aspronisi, a remnant of a much larger island. The main port is at **Korfos** although day trips from **Thira** land at **Riva** and visitors ascend to the caldera rim. **Manolas**, the main town high above **Korfos**, still has its windmill intact. **Nea Kamena** has hot springs and an active crater. Both islands can be visited on day trips from **Skala Thira** or **Ammoudhi** on Thira.

Below: *Crystal clear seas and guaranteed sunshine make Ios a tourist idyll.*

A VOLCANIC TOWN

The first inhabitants of **Thira** (from **Karia**) were ousted by the Minoans whose extensive colony at **Akrotiri** was destroyed by a volcanic eruption in 1450BC. In 1967 the archaeologist Marinatos began to dig at Akrotiri where **Minoan** vases had been found. The discovery of a complete Minoan colony buried in volcanic *tephra* complete with marvellous frescoes and ceramics was the find of the century.

Thira, the island capital, teeters high on the crater rim, above **Skala**, where ships berth. Nearly 600 steps lead from sea level to the village; alternatives to walking include renting a mule or taking the cable car to the outskirts of Thira. The village is highly commercialized (and dedicated to fleecing a steady flow of day visitors from cruise ships) but its **Archaeological Museum**, with a display of Cycladic figurines and finds from the Akrotiri site, is worth a look.

Firostefani and **Imerovigli** villages, just north of Thira, have become virtual suburbs of the capital, but **Oia**, on the north tip of the island, is one of the most enchanting spots in the Aegean. Almost destroyed by an earthquake in 1956, it has been rebuilt since the 1990s and now boasts an array of great restaurants and cafés, chic boutiques, and some of Greece's most luxurious small hotels.

Anafi ★

Anafi is the end of the line. Few ferries stop here, none travel onward, and tourism barely exists – apart from a hardy handful of solitude seekers who, having discovered its simple charms (including lovely if tiny beaches) come back every year. **Chora**, the island's one village, is a short but steep walk from the jetty at **Agios Nikolaos**. A path leads eastward through deserted **Kalamitsa** to the **Monastery of Panagia Kalamiotissa**, on the eastern tip of the island.

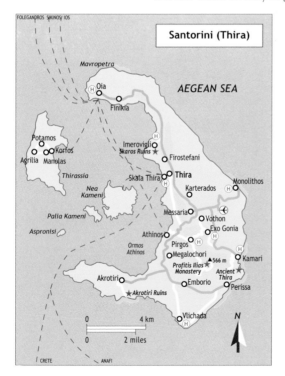

Santorini (Thira)

THE WESTERN CYCLADES

The western Cyclades are much less visited – partly because few ferries connect them with their eastern neighbours – but have plenty to recommend them.

Kea (Tzea) ★★

Within sight of Cape Sounion, the southern tip of Attica, Kea (area 131km²; 51 sq miles) is popular with Athenian weekenders and

sees few foreigners. Those who disembark at **Korissia**, on the deep natural harbour of **Agios Nikolaos Bay**, are rewarded by pretty landscapes of terraced hills and fertile valleys, peppered with red pantiled houses, windmills and tiny churches. **Ioulis** (Chora) the surprisingly cosmopolitan main village, is about 6km (4 miles) from the port, set between two hills. The site of **ancient Ioulis** – whose citizens were traditionally expected to end their lives at 70 with a lethal draught of hemlock to avoid becoming a burden on the state – is buried beneath the ruins of a Venetian castle built in 1210. To the northeast of the town is the sphinx-like **Lion of Kea**, carved in the 6th century BC from a rocky outcrop. Kea is a good island for walkers, and there are good beaches at **Pisses**, **Vourkari** and **Koundouros** on the west coast and **Otzias** in the north.

Above: *Dark volcanic cliffs contrast dramatically with shining white Thira, capital of the island of Santorini (Thira).*

Kythnos ★

Barren and mountainous Kythnos (area 86km²; 33 sq miles) has a jagged coastline of small, rocky bays, with one small beach on the east coast at **Agios Stefanos**, within walking distance of its Chora. Ferries dock at **Merinas**, 12 km (7.5 miles) from Chora, an attractive, typically Cycladic town with two pretty churches, **Agios Savvas** and **Agia Triadha**. **Driopida**, once the main village, stands on a ridge in the centre of the island. It was once famed for its ceramics and in the labyrinthine **Galatas quarter** there is a pottery workshop. The mildly radioactive thermal spring at **Loutra** is popular with elderly Greeks seeking cures for rheumatism and arthritis, and the oldest known habitation in the Cyclades (7500–6000BC) has been found nearby.

A PORT IN A STORM

According to Greek legend, **Anafi** was created by Apollo for Jason and the Argonauts in need of a no-frills berth during a storm. A **Temple to Apollo** later stood on the island, built by a grateful Jason, but its stone was apparently used to build a monastery – Zoodochos Pigi.

PAROS

Ormos Agios
Georgiou

Herortissos

SERIFOS

Agios
Marina

Sifnos

0   3 km

0   1 mile

Artemonas

Kamares   Kato Petali
                 Kastro

Apollonia
          Exambela   Ormos
                            Seralia
Katavati

Agios
Andreas ★
Vathi   Platis   Faros
          Gialos   Chrissopigi
                    Monastery

MILOS

Kitriani   N

AEGEAN SEA

KIMOLOS

Serifos ★★★

Serifos is the most enchanting and peaceful of the western Cyclades, with uncrowded sandy beaches, a charming mountain-top Chora beneath a crumbling Venetian castle, and fine island walking trails leading through terraced valleys and farming villages. Many visitors stray no further than the pretty harbour village, **Livadi**, simply because most of the accommodation is here, along with places to eat and a good, clean sandy beach. For those who want to explore further afield, there are lovely sandy bays at **Agios Sostis**, **Chalara**, **Gialos**, **Gianema**, **Lia, Platys** and **Psili Ammos** around Serifos's roughly circular coastline; there are simple beach restaurants on most of these in summer.

Sifnos ★★★

Rising from the sea, the terraced hills of Sifnos are dotted with villages and tiny chapels, old stone towers and whitewashed Venetian dovecotes. **Apollonia**, the sleepy capital, stands high above the sea, hidden from the little port of **Kamares** on the west coast, and is one of the most attractive island capitals. It has escaped the worst of the backpacker invasions and of package tourism, and attracts a loyal and discerning clientele, with some fine hotels and excellent traditional restaurants. Its name comes from the ancient **Temple of Apollo** which now lies beneath the 18th-century

Church of Panagia Ouranofora, which was partly built with stones from the temple.

Sifnos is famous for its pottery, and examples of traditional work can be seen in the **Museum of Popular Arts and Folklore** (open daily 10:00–13:00, 18:00–22:00), along with fine embroidery and island costumes. Several potters work in the village and sell ceramics influenced by island traditions, as well as more modern work. Sifnos's beaches are in the southern part of the island, at **Vathi**, **Platis Gialos** and **Faros**, and the **Monastery of Chrissopigi** stands on a headland between **Platis Gialos** and **Faros**.

Milos ★★

Milos's volcanic coast and hinterland are carved into weird shapes by sea, wind and human activity. Its hills are scarred by china clay (kaolin) quarrying and its wind-sculpted cliffs, steaming hot springs and sulphurous mineral deposits give the island an other-worldly look. Milos is cut almost in two by a deep bay on which stands its port, **Adamas**. **Plaka**, the main village, is a whitewashed chora which is prettiest at sunset and best seen from the two chapels (Panagia Skiniotissa and Panagia Thalassitras) on the hill above. Most of the islanders live in the surrounding villages of **Plakes**, **Pera**, **Triovasalos** and **Tripiti**, and an old cobbled mule track leads to **Klima**, a marvellously photogenic village with brightly painted boathouses. Covering about 161km² (62 sq miles), Milos is not known for its beaches, but the waters round its rocky coast are very clear and offer excellent snorkelling.

Kimolos ★

There is barely room for the ferry to squeeze through the narrow strait between Kimolos and Milos, its larger neighbour. Kimolos is small (38km²; 15 sq miles) and its northern half has been heavily quarried, but its little-visited south may appeal to those in search of the real Greece. Psathi, the port, is set below the main village, a tumbledown maze of old houses in and around the ramparts of a Venetian castle, with a few tavernas and places to stay in Exo Kastro, outside the walls.

64

The Cyclades at a Glance

BEST TIMES TO VISIT

The Cyclades tourist season starts at Easter and is busiest Jun–Sep. Visitors come to Paros, Mykonos and Santorini till end Oct (Mykonos all year round). Elsewhere, hotels, guesthouses and restaurants close from late Oct–Apr and infrequent ferry schedules make winter travel difficult.

GETTING THERE

By Air: Olympic Airlines flies from Athens to Mykonos, Santorini, Milos, Naxos and Paros. Aegean Airlines flies to Mykonos, Santorini. Summer charter flights from London and other UK airports and from some mainland European cities go to Mykonos and Santorini.
Olympic Airlines, tel: Athens 21092 69111, www.olympicairlines.com
Aegean Airlines, call centre tel: (for all flights) 80111 20000, www.aegeanair.com
By Sea: Frequent ferries and catamarans link the Cyclades with Rafina, Piraeus, Crete and the Dodecanese. Schedules change frequently – the most reliable sources of information are the local port authority and the essential Greek Travel Pages: www.gtp.gr There are three main ferry routes: from Piraeus through the western Cyclades, ending in Milos or Folegandros; from Piraeus or Rafina through the eastern Cyclades, via Andros-Tinos-Mykonos-Paros-Naxos and Santorini or Amorgos; and

eastward via Mykonos to Ikaria, Fourni and Samos. Ferries also link Mykonos, Amorgos and Paros with Kos and Rhodes. Off these main routes, travel between islands can be frustrating (it can take a full day to get to Amorgos from Santorini, changing at Naxos). There are also ferries from Lavrio, on the Attica peninsula, to Kea and Kythnos. Ferry companies include: **Anafi-Santorini Lines**, tel: 22860 61203 (Anafi-Santorini). **Blue Star Ferries**, tel: 21089 19800, www.bluestar ferries.com **C-Link Ferries**, tel: 21045 94096, (Piraeus-Andros, Tinos). **Hellenic Seaways**, tel: 21041 99000, www.hellenic seaways.gr (Zea to Andros, Tinos, Paros, Mykonos, Syros, Santorini, Amorgos, Kea). **Karystos Shipping**, tel: 21042 20548 (Lavrios-Kea). **GA Ferries**, tel: 21045 11720 (Piraeus to all major Cyclades and Crete). **L.A.N.E.**, tel: 21042 74011 (Crete-Milos). **Minoan Lines**, tel: 28103 99800, www.minoan.gr (Piraeus to all major Cyclades and Crete).

WHERE TO STAY

Booking is advisable for Greek Easter and August holidays, and also for the best hotels in Mykonos and Santorini. The latter has a reputation for some of the world's most stylish 'boutique hotels'. Contact Small Luxury Hotels (www.slh.com).
Mykonos
LUXURY
Mykonos Theoxenia, Kato Mili,

tel: 22890 22230, fax: 22890 23008. Stylish, with a big pool, sea views and within a short walk of the picturesque Chora and harbour.
Hotel Belvedere, Rochari, Mykonos Town, tel: 22890 025122, fax: 22890 025126. An 18th-century mansion; pool, super courtyard restaurant; magnificent wine cellar.
Kivotos Clubhotel, Ornos Bay, tel: 22890 24094, fax: 22890 22844. Exclusive and romantic; private beach.
Semeli, Rochari, Mykonos Town, tel: 22890 27466, fax: 22890 27467. Rooms furnished with antiques in a village stately home; pool.
MID-RANGE
Hotel Elena, Rochari, Mykonos Town, tel: 22890 22361, fax: 22890 24112. Rooms and apartments, wooden balconies.
BUDGET
Hotel Matina, Fournakia 3, Mykonos Town, tel: 22890 22387, fax: 22890 24501. Good value, with shaded garden in the heart of the Chora.
Santorini
LUXURY
Chromata, Imerovigli, tel: 22860 24850, fax: 22860 23278. Colourful and hip hotel; suites with whirlpool baths.
Kastelli Resort, Kamari, tel: 22860 31530, fax: 22860 32530. Two large pools, two restaurants, tennis courts, vineyard. Near Kamari beach.
Katikies, Oia, tel: 22860 71401, fax: 22860 71129. Magnificent sunset views and

The Cyclades at a Glance

a fine gourmet restaurant.

Notos Therme & Spa,
Vlichada, tel: 22860 81115,
fax: 22860 81266. Superb
thalassotherapy resort on the
beach, with elegant terraces
cascading down to the sea.

Zannos Melathron, Pirgos,
tel: 22860 28220. Designer
mansion with panoramic
views, huge suites, a pool,
fine wining and dining and
attentive staff.

MID-RANGE

Rose Bay, Kamari, tel: 22860
33650, fax: 22860 33653.
Colourful and comfortable,
with a pool, at Kamari beach.

BUDGET

Hotel Fregata, Oia, tel: 22860
71221, fax: 22860 71333.
Simple hotel with the same fine
view as its luxury neighbours at
less than a quarter of the price.

Amorgos

MID-RANGE/BUDGET

Lakki Village, Aigiali, tel: 22850
73253, fax: 22850 73244.
Simple rooms, clean apartments and cottages on a sandy
beach, near Aigiali village.

WHERE TO EAT

Eating places genuinely worth
visiting include:

Mykonos

Belvedere Sushi, Hotel
Belvedere, Mykonos Town,
tel: 22890 25122. Not cheap,
but serves wonderful sushi
and good cocktails beside
the hotel's pool.

Diles, Hotel Andromeda, Platia
Lakka, Mykonos Town, tel:
22890 22120. Highly rated

luxury restaurant, also offers
fixed-price 'economy' menus.

Santorini

Archipelagos, Thira Town, tel:
22860 23673. Serves Greek
and international dishes with a
fine view of the caldera.

Iliovasilema, Ammoudi, Oia,
tel: 22860 71614. Excellent
fish restaurant, great value.

1800, Oia, tel: 22860 71485.
This exclusive (and expensive)
restaurant serves 'new
Mediterranean' cuisine.

Pelekanos Café, Oia, tel:
22860 71553. Friendly and
unpretentious, light meals and
drinks with a great sunset view
from the rooftop terrace.

TOURS AND EXCURSIONS

From Mykonos the most
popular excursion is to Delos
(45 min by boat). First boats
leave around 08:00 and
operate every couple of hours
through the day, with the last
leaving Delos around 18:00.
You can visit Tinos for the day
by ferry. Scuba diving trips are
on offer from Mykonos and
Santorini, where other
popular excursions include
cruises to the volcano islands
and Therasia and sunset
cruises in the caldera.

USEFUL CONTACTS

Websites:
www.cyclades-tour.gr
www.thegreektravel.com
www.naxos-greece.net

Port authorities:

Andros, tel: 22820 22250.
Tinos, tel: 22830 22348.
Siros, tel: 22810 88888.
Mykonos, tel: 22890 22218.
Paros, tel: 22840 21240.
Naxos, tel: 22850 22300.
Amorgos, tel: 22850 71259.
Folegandros, tel: 22860
41249. **Ios**, tel: 22860 91264.
Santorini, tel: 22860 22239.
Kea, tel: 22920 21344.
Kythnos, tel: 22810 31290.
Kimolos, tel: 22870 22100.
Milos, tel: 22870 22155.
Serifos, tel: 22810 51470.
Sifnos, tel: 22840 33617.
Tourist offices: Andros, tel:
22820 24860. Santorini, tel:
22860 27155.
Tourist Police: Andros, tel:
22820 22300. **Tinos**, tel: 22830
23670. **Siros**, tel: 22810 96123.
Mykonos, tel: 22890 22218.
Paros, tel: 22840 23333.
Naxos, tel: 22850 22100.
Police: Amorgos, tel: 22850
71210. **Folegandros**, tel: 22860
41249. **Ios**, tel: 22860 91222.
Santorini, tel: 22860 22649.

CYCLADES	J	F	M	A	M	J	J	A	S	O	N	D
AVERAGE TEMP. °F	52	52	55	61	68	75	79	77	73	66	59	54
AVERAGE TEMP. °C	11	11	13	16	20	24	26	25	23	19	15	12
HOURS OF SUN DAILY	5	5	6	7	10	11	12	11	9	7	5	4
RAINFALL ins.	2.5	2	2	1	0.5	0	0	0	0	1.5	2	3
RAINFALL mm	66	52	47	18	10	3	1	2	6	36	47	70
DAYS OF RAINFALL	14	11	10	7	4	1	1	1	2	6	9	14

5
Crete

For many visitors, Crete is the logical end of an island-hopping trip, with the option of travelling back to Athens by sea or air after exploring the island's mountains, relaxing on its beaches, or viewing its outstanding archaeological sites, medieval castles and museums. Thousands more fly straight to the island's two airports – **Hania** and **Irakliou** – from Europe for sun, sand and sightseeing.

The largest of the Greek islands, Crete is almost a country in its own right. Some 227km (173 miles) from east to west, it is dominated by mountain ranges that rise to peaks of more than 2400m (7847ft) and remain snow-capped as late as June. It has its own dialect, customs, and folklore, and joined the modern Greek nation only in 1913 after a long struggle to throw off the Ottoman yoke.

Running through Crete's barren mountains are deep canyons, and around its coasts as well as in hidden inland valleys are pockets of richly fertile land where fruit and vegetables grow in profusion. Vineyards on the northern slopes produce some of Greece's best wines, and the fertile Messara plain in the south is covered with greenhouses producing vegetables for export almost year round.

Tourism extends its reach to every corner of the island. The north coast – from **Agios Nikolaos** in the east to **Hania** in the west – is the most intensively developed region, with luxury hotel complexes in the Elounda region, downmarket resorts dedicated to alcohol-fuelled excess at **Malia** and **Hersonissos**, and better-managed, mid-range tourism focused on the attractive Venetian harbour towns of **Hania** and **Rethimnon**.

MEDITERRANEAN SEA

DON'T MISS

*** Knossos: remains of an ancient city including Minoan palace with Sir Arthur Evans' painted reconstructions.
*** Irakliou Archaeology Museum: an outstanding museum by any standards.
** Samaria Gorge: walk from mountains to the sea in Europe's longest gorge.
** Phaestos and Agia Triadha: important archaeological site for Minoan grandeur in great scenery.

Opposite: *Knossos, most famous of the Minoan palaces, can be relatively crowd free in the early mornings or late evenings.*

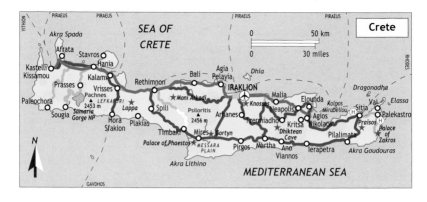

During Crete's brief, colourful spring, swathes of wild flowers cover every hillside and scrap of waste land. The Cretan struggle against foreign occupation – by Venetians during the Middle Ages, Ottoman Turks from the 17th–20th centuries, and German forces during World War II – is the stuff of local legend, and Cretans are still fiercely proud of their distinctive island culture. Though only a few very old men now wear the mountaineer's costume of black boots, baggy trousers, black waistcoat and headscarf, traditional music and dance are still very much alive.

Crete is divided into four provinces (Nomes). From west to east: **Hania**, **Rethimnon**, **Irakliou** and **Lasithiou**, each of which spans a slice of the north and south coasts.

NORTH COAST
Nomos Irakliou ★★

Irakliou, the fifth largest city in Greece and the Cretan capital since 1971, was first established as the Minoan port of Knossos and later became the Saracen base for piracy and the slave trade. An exploration of old Irakliou can begin with the Venetian harbour: the **Lion of St Mark** guards the restored **Rocco al Mare**, a 16th-century fortress; the nearby arches of the **Venetian Arsenali** (the shipyards) are partially hidden by the street.

CLIMATE

A southerly position gives Crete a dry Mediterranean climate – the northern coast seems lush in comparison with the south which is arid. Winter rains fall on the high mountains as snow, which persists well into the summer on the heights.

Below: *Reconstruction of Minoan frescoes on inner walls at Knossos.*

The busy main thoroughfare, 25 Avgoustou, leads away from the harbour towards the **Venetian Loggia** (a meeting place for the nobility) and the **Venetian Church of San Marco** (1239); the **Morosins Fountain** with its lions stands in Platia Venizelou. By law, the **Archaeological Museum** near Platia Eleftherias can claim every important artefact found on the island, thus its Minoan collections have no equal and it is definitely worth a visit.

The island of Dhia just off the coast offers a day's escape from Irakliou: the island is a reserve for **kri-kri** (*Agrimi*), the Cretan wild goat (*see* p. 10).

Knossos ★★★

Knossos, the second most visited site in Greece after the Acropolis, is served every 10 minutes by buses from the modern harbour. Visit as it opens or out of season to avoid crowds drawn by the fascinating but sometimes fanciful reconstructions by Sir Arthur Evans (1900–1920). Visitors can see remains of a magnificent **Minoan** palace, royal villa and caravanserai from the Neopalatial period (1700–1380BC). The city of Knossos housed a population of between 30,000 and 100,000; the last palace to be built apparently had over 1300 rooms and was perhaps the inspiration for the myth of the 'labyrinth' in which Theseus fought the Minotaur. Open daily 08:00–19:30 mid-April to end Oct, 08:00–17:30 Nov to Easter.

Nomos Lasithiou – Eastern Crete ★★

Agios Nikolaos, now Crete's best-known harbour resort, was once the harbour for the city of Lato. Kritsa has

Above: *Once a sleepy harbour, Agios Nikolaos is now a busy town.*

WHITE MOUNTAINS

Entrance to the Samaria Gorge lies at the southern end of the Omalos Plateau. A fee is charged which provides for wildlife rangers before descent of the steep but well-kept Xiloskalo (wooden staircase). From the gorge entrance the walk to Agia Roumeli takes five to eight hours depending on slopes and pace. The path passes the tiny chapel of Samaria and continues southwards to the coast through the Sideroportes ('Iron Gates') where the gorge narrows dramatically. There are several hotels at Agia Roumeli reached only by ferries which operate to Paleochora and Hora Sfakion. Over 250,000 people visit annually but their route is restricted to the bottom of the gorge. Numerous flowers appear on the rock walls in spring.

Above: *Rethimnon harbour retains a great deal of charm and some good fish restaurants.*

MINOAN CIVILIZATION

The Copper and Bronze Age Minoan civilization falls into distinct periods –
Pre-Palatial 2600–1900BC: sanctuaries built in high places, first monumental *tholos* tombs.
Old Palace Period 1900–1700BC: first plumbing, bull culture predominant, Crete dominated the seas, thalassocracy.
New Palace Period 1700–1450BC: elaborate rebuilding of palaces and fancy villas after devastating earthquake, art flourishes.
Post-Palace Period 1450: volcanic eruption hits Thira (Santorini).
1450–1100BC: tidal waves and earthquakes hit Crete, gradual infiltration of the Mycenaean culture.

become the traditional Cretan village for day trips from Agios Nikolaos as its 13th-century church (Panayia Kera) has superb Byzantine frescoes. North of Kritsa lies Lato, a 4km (2½ miles) drive, lying between two hills. The corniche road along the **Gulf of Mirabello** leads to **Sitia**, passing several noteworthy sites: **Vronda**, a Minoan cemetery, and further on the lofty **Kastro** with superb views over the gulf. Sitia with its long, sandy beach has retained its soul compared with other resorts.

The surrounding countryside has a wealth of archaeological sites. **Vai**, a tourist magnet, boasts a superb beach and grove of palm trees (*Phoenix theophrastus*) of a species unique to Crete. The fertile **Lasithiou Plateau** with its thousands of windmills (no longer with sails) can be reached by buses from Malia, Irakliou or Agios Nikolaos. From Psychro on the western edge of the plain, the climb begins to the Dhiktean cave – birthplace of Zeus and a shrine from Minoan times.

Nomos Rethimnon – Central Crete ★★

Rethimnon's delightful Venetian harbour is surrounded by fish restaurants and a pair of minarets add an eastern touch to the skyline – one above the **Nerandzes Mosque,** now a museum, (open daily 11:00–19:00, closed August) offers views over the old quarter. Near the harbour is the **Loggia**, built in 1600. Finest of the Venetian buildings, it is now the city library. The acropolis of ancient Rethimnon lies below the massive Venetian castle (Fortezza); the **Archaeological Museum** at its entrance was formerly a Turkish prison.

The huge bulk of snow-capped **Psiloritis** (Mt Idha) soars to 2456m (8058ft) some 25km (16 miles) southwest of Rethimnon. There are excellent walks (organized treks are advertised in Hania) and the wild **Amari Valley** on the southwest flank is a favourite with birdwatchers for its bearded vultures. **Arkadi Monastery**, set in wild country on the northern slopes, looks like a small fort.

Nomos Hania – Western Crete ★★

Hania, once the Turkish capital, is a great place to wander and find remnants of its **Venetian** and **Ottoman** buildings. Hania's harbour and covered market are popular. The impressively domed **Mosque of the Janissaries** lies just off the harbour; **Kastelli**, the hill behind it, has been occupied since Neolithic times. The **White Mountains** (Lefka Ori) reach 2453m (8048ft) at **Mt Pachnes** and are cut by deep ravines – the most famous is the **Samaria Gorge** which, at 18km (11 miles), is the longest in Europe.

West of Hania, the two peninsulas of Rodhopou and Gramvoussa are crowd free and scenically striking. A minor road leads off the main coast road as far as **Afrata**, last village on the **Rodhopou** peninsula, and a track leads for some 12km (7½ miles) towards the tip. Similarly, the **Gramvoussa** peninsula has to be explored on tracks – either on foot, by moped or by boat from Hania harbour.

AKROTIRI

Akrotiri, the promontory to the east of Hania, makes a worthwhile day trip. At the monastery of Agia Triadha tangerine trees surround its Venetian façade; the monastery of Moni Gouvernetou stands on a plateau 5km (3 miles) further north. A rough path leads from the latter down a rocky gorge spanned by a bridge going nowhere, and also to the monastery of Moni Katholiko which is built into the rocks.

SOUTH COAST

The wild, unspoiled southern coast east and west of **Hora Sfakion** is green in winter and spring but arid the rest of the year. Two gorges, smaller than Samaria and less popular, can be reached from Hora Sfakion: the **Imbros Gorge** lies below the road as it descends to Hora Sfakion from the **Askifou Plateau**. The **Aradhena Gorge** begins from the deserted village of **Aradhena**, 4km (2½ miles) from Anopoli.

From Irakliou the road crosses the mountains and descends with panoramic views over the **Messara Plain**. Following the Dorian invasion in 1100BC, **Gortyn** became an important city as the Roman capital of Cyrenaica (Crete and Libya). The **Law Code of Gortyn** is built into the wall of the **Roman Odeion** and the nearby **Christian Basilica of Agios Titos** (AD105) is the best preserved in Crete.

The Minoan palace of **Phaestos** lies on a hill close to **Mires** with superb views over the plain to the mountains beyond (open daily 08:00–20:00 in summer, 08:00–19:00 in winter). A road runs 3km (2 miles) east along the ridge to **Agia Triadha** – once possibly a Minoan royal summer palace (open daily 08:00–19:00/20:00 in summer; 08:00–15:00 in winter). The site is incomparable.

Below: *Home of rare plants and birds, Samaria Gorge offers one of the most dramatic walks in Europe.*

Crete at a Glance

Spring flowers bloom Mar–Apr and many villages hold Easter processions. May–Jul is best for mountain flowers but Samaria Gorge is closed until 1 May. Beaches are crowded from Jun–Sep but by Oct they empty (the sea is still warm). Winters are mild.

GB Airways has direct flights to Irakliou from London Gatwick (2 per week). Olympic Airways and Aegean Airlines fly to Irakliou and Hania from Athens and Thessaloniki; there are several services weekly between Irakliou and Rhodes. In season there are many **charter flights** to both airports from UK and other European countries. **Irakliou** airport is connected with the capital by bus #1 to **Platia Eleftherias**. In **Hania**, Olympic buses meet the company flights and many hotels have their own coaches. **Taxis** are quick and reliable. There are daily **ferries** from **Piraeus** to both **Irakliou** and **Hania**. At **Hania**, ferries dock at **Soudha**, from where buses and taxis take you into town. You can also travel direct from the Cyclades and the Dodecanese (see p. 124 for on-line timetable and bookings).

Frequent KTEL buses operate along the north coast – Kastelli, Hania and Rethimnon connect with Irakliou, Agios Nikolaos and Sitia. Other bus routes are between Irakliou and Matala and Agia Galini on the south coast; Sitia and Ierapetra; and Hania and Paleochora. Long-distance, fixed-price taxis are easy to find and affordable. Small coastal ferries go between Paleochora and Chora Sfakion, calling at Sougia, Agia Roumeli and Loutro. Ferry lines include: **ANEK Lines**, Piraeus, tel: 21041 97420; Hania, tel: 28210 27500; Irakliou, tel: 28102 22481; Rethimnon, tel: 28310 29874. **Blue Star Ferries**, tel: 21089 19950, www.bluestarferries.gr **GA Ferries**, tel: 21045 82640. **Minoan Lines**, tel: Piraeus 21040 82480, Irakliou, tel: 28102 29602, www.minoan.gr

Hania
LUXURY
Amphora, 20 Parados Theotokopoulou St., tel: 28210 93226, fax: 28210 93224. Lovely 14th-century building on harbour front, nice rooms.
MID-RANGE
Pension Nostos, 42–46 Zambeliou, 73100, tel: 28210 94743, fax: 28210 94740. Pension with 12 studios.
Doma, Venizelou 124, tel: 28210 51772, fax: 28210 41578. Formerly the Austrian Consulate, comfortable, with antiques and memorabilia.
BUDGET
Pension Eva, 1 Theofanous and Zambeliou St., 73100,

tel: 28210 76706. Immaculate period-style rooms.
Irakliou
LUXURY
Astoria Capsis Hotel, Platia Eleftherias, tel: 28103 43080, fax: 28102 92078. Central.
MID-RANGE
Mediterranean, Smirnis St., tel: 28102 89331, fax: 21802 89335. Reasonable, central.
BUDGET
Olympic, Platia Kournarou, tel: 28210 288861, fax: 28210 22512. Town centre.
Rethimnon
LUXURY
Theartemis Palace, Portaliou 30, tel: 28310 53991, fax: 28310 23785. Modern, central, with a courtyard pool.
MID-RANGE
Mythos Suites, Karaoli 12 (corner of Dimitriou), tel: 28310 53917, fax: 28310 51306. Delightful small hotel with 14 suites and rooms around a courtyard with a tiny pool.
BUDGET
Costis Apartments, Kondylaki 12, tel: 28310 29159, fax: 28310 51802. Very good value with pool and garden close to the town centre and 600m from the beach.
Hersonissos
LUXURY
Aldemar Knossos Royal Village, Hersonissos 70014, tel: 28970 27400, fax: 28970 23150. Beach location, pool.
MID-RANGE
Maria Apartments, Hersonissos 70014, tel: 28970 22580. Spacious apartments.

Crete at a Glance

Budget
Hotel Iro, Hersonissos 70014, tel: 28970 22136, fax: 28970 23728. Central.

Malia
Budget
Pension Aspasia, Malia 70007, tel: 28970 31290. Clean, friendly pension just outside the old village.

Matala
Luxury
The Valley Village, tel: 28920 45776. On the edge of village, with pool.
Mid-range
Hotel Zafiria, tel: 28920 45112, fax: 28920 45747. Reasonably priced, near town.

Agios Nikolaos
Luxury
Elounda Mare, tel: 28410 25041, fax: 28410 41307. Superb complex of hotel rooms, 40 bungalows.
St Nicolas Bay, tel: 28410 25041. Complex with 130 bungalows.
Minos Beach Art'otel, Ag. Nikolaos, tel: 28410 22345, fax: 28410 22548. Magnificent resort village with a superb pool and four fine restaurants
Mid-range
Panorama, Akti Koundourou St., tel: 28410 28890. Good views over harbour.

Hania
O Dinos, Harbour. Seafood dishes, reasonably priced.
Tamam, 49 Zambeliou Street. Greek dishes, vegetarian food.

Kings, 15 Kondilaki Street. International cusine, seafood and vegetarian specialities.
Konaki, 40 Kondilaki St., tel: 02821 97130. Cretan specialites in old courtyard.

Irakliou
Gao, Platia Venizelou, tel: 28210 46338. Self-service cafeteria, pasta dishes.
Giovanni, Korai St., tel: 28103 46338. Traditional Cretan dishes and vegetarian menu.

Rethimnon
Avli, Xanthoudido 22, tel: 28310 26213. Best restaurant in northern Crete, with a fine wine list.

Hersonissos
La Fontanina, Georgiou Petraki St., tel: 28970 22209. Excellent pizzeria/pasta house.
Sokaki, 10 Evagelistias St., tel: 28970 23972. Cretan special-ities at the right price.

Malia
San Georgio, on main square, tel: 28970 32211. Greek food.

Agios Nikolaos
Il Capriccio, 31 Akti Koundourou. Italian menu.

Coach tours to **Knossos**, **Festos** and **Agia Triadha**, **Samaria** and **Vai** organized by travel agents in main towns. Sea cruises to Siss, Mohlos and Psira islands (from **Agios Nikolaos**) and Dhia (Irakliou). Hellenic Mountaineering Society organizes trekking in the White Mountains, Psiloritis, around the island, tel: 28250 44946 and 28310 55855.

Tourist Office:
Irakliou, 1 Xanthoudidou Street, tel: 28102 28203.
Hania, 18 Kriari Street, tel: 28210 92943.
Rethimnon, E, Venizelou, tel: 28310 29148.
Agios Nikolaos, Agios Nikolaos Marina, tel: 28410 82384, fax: (0841) 26-398. For information in English, tel: 131.
Tourist Police:
Irakliou, Dikeosinos Street, tel: 28102 83190.
Hania, 40 Kriari Street, tel: 28210 28708.
Rethimnon, Platia Heroon, tel: 28310 28156.
Agios Nikolaos, tel: 28410 26900.

Websites:
www.cretetravel.com
www.west-crete.com

CRETE	J	F	M	A	M	J	J	A	S	O	N	D
AVERAGE TEMP. °F	54	54	57	63	68	75	79	79	75	70	64	57
AVERAGE TEMP. °C	12	12	14	17	20	24	26	26	24	21	18	14
HOURS OF SUN DAILY	5	5	6	8	10	12	13	12	9	6	5	5
RAINFALL ins.	15	15	17	20	24	29	31	31	29	25	20	17
RAINFALL mm	5	5	6	8	10	12	13	12	9	6	5	5
DAYS OF RAINFALL	9	8	6	3	1	0	0	0	1	3	6	9

6
The Dodecanese

Closer to the Asia Minor coast than to the Greek mainland, and last of the islands to rejoin modern Greece, the Dodecanese have a distinctive character and architecture that sets them apart from their western neighbours in the Cyclades. Their name means 'twelve islands' – but the group now comprises 17 inhabited islands, large and small, and at least as many more desert islets. Their landscapes range from the barren limestone hills of **Tilos** and **Symi** to the lush vineyards of **Rhodes**, the fertile slopes of **Kos** and the bubbling volcanic crater of **Nissiros**.

Over the last seven centuries, the Dodecanese have been ruled variously by piratical Genoese nobles, the holy warriors of the Order of St John, Ottoman pashas and Italian colonialists. On most of the islands, local architecture reflects that history. Brooding medieval castles overlook the tumbled columns of Hellenistic temples, abandoned mosques and hamams stand beside Italian Art Deco buildings, and pantiled houses with wrought-iron balconies lean over island harbours.

Rhodes and **Kos**, with their long beaches and longer summers, along with international airports, attract package holiday-makers by the million, and package tourism spills over onto much smaller neighbouring islands like **Chalki**, **Symi**, and **Kalimnos**. **Karpathos** and **Leros** also have airports but are far less developed than the big two. The smaller islands at the north and south ends of the Dodecanese chain, as well as lonely **Astipalea**, are perfect for would-be castaways. With several ferries daily plying up and down the chain (and onward to Crete from

MEDITERRANEAN SEA

Opposite: *Lindos, where white houses ring the acropolis, is one of the best-loved places in Greece.*

Rhodes and to the Northeast Aegean isles and the Cyclades from Kos) the Dodecanese are also great for island hopping.

RHODES ★★★

With 300 days of sunshine a year, a magnificent medieval walled town that is listed as a World Heritage Site, and some excellent beaches, the largest island in the Dodecanese chain (1398km²; 540 sq miles) is understandably popular. Rhodes's big resorts cluster either side of Rhodes town, with a second area of intense tourism development north and south of Lindos, midway along the east coast. The island's mountainous and windswept west coast is much less developed, but also much less appealing.

Rhodes town is three towns in one. The **Old Town**, built by the Knights of St John, is a medieval marvel, with battlements and bastions enclosing a labyrinth of narrow streets and stone houses that gets its fair share of tourism but is still a living, working community. Outside the walls, the **New Town** is a lively, cosmopolitan miniature of Athens where most Rhodians live, work and play. Blending into its outskirts are the island's busiest resort areas – **Ixia**, **Trianda** and **Ialyssos** on the west coast and, a little further from town on the east coast, **Koskinou**, **Faliraki** and **Afandou**. All of these offer luxurious

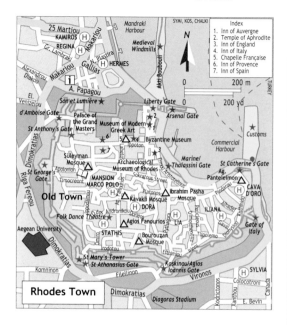

Rhodes Town

Index
1. Inn of Auvergne
2. Temple of Aphrodite
3. Inn of England
4. Inn of Italy
5. Chapelle Française
6. Inn of Provence
7. Inn of Spain

resort hotels with pools, international wining and dining,
and lively nightlife – perhaps a little too lively in the case
of **Faliraki**, which is struggling to shed a reputation for
drunken misbehaviour by young Britons.

The Old Town is dominated by the magnificent
Palace of the Grand Masters (open 12:30–20:00
Mon, 08:00–20:00 Tue–Sun), rebuilt during the
Italian occupation of the island (1912–1943) and
standing at the head of the **Street of the Knights**,
where eight **Inns** housed each of the national chap-
ters within the order. The original **Hospital of the
Knights** is now the **Archaeological Museum**. At
Platia Simis, the **Museum of Modern Greek Art**
houses 20th-century engraving, sculpture, and old
maps and prints; most of the paintings in its perma-
nent collection are now in the new **Nestoridio
Melathro Gallery** on Platia Haritous in the New
Town. Several mosques (currently under restoration)
are relics of Turkish rule and on the **Platia Evreon**

Below: *Twin towers
dominate the Palace
of the Grand Masters.*

VALLEY OF THE BUTTERFLIES

Petaloudes, the 'Valley of the Butterflies' on Rhodes, is a summer attraction where **Jersey tiger moths** (*Euplagia quadripunctaria*) – not butterflies – gather in their thousands in the cool, damp valley. Visitors are warned not to disturb the creatures as their numbers decrease annually but some people are tempted to put them to flight (the flash of bright orange under wings is spectacular). The valley is well signposted west of Maritsa, off the road from Psinthos to Kalamonas. In recent years fires have damaged the woods.

LINDOS'S BEACHES

Although Lindos beach is crowded, the bays south of **Lardos** are less frequented. The coast north of Lindos offers a number of good but busy beaches from **Kalithea** to **Faliraki** – the island's major resort featuring every watersport imaginable for the 18–30 set. **Afandou** is slightly quieter and has an 18-hole golf course; south of **Kolimbia** the beaches near **Vagia point** are still unspoiled.

Martyron (Square of the Martyrs) a monument commemorates the centuries-old Jewish community annihilated during the German occupation of 1943–45.

The **Folk Dance Theatre**, on Odos Andronikou holds authentic Greek dance performances (Mon, Wed and Fri, May–Oct).

Classical relics include a few columns of the **Temple of Pythian Apollo** on 'Monte Smith' (named after a 19th-century British admiral) 2km (1.5 miles) west of town, and an **ancient theatre** – all that remains of the Hellenistic city. Remains of other classical city-states can be seen at ancient **Ialyssos** above the modern resort of the same name, and at ancient **Kamiros** and on the acropolis of **Lindos**, above the island's prettiest village. Modern Lindos is a photogenic jumble of geometric, whitewashed houses with pebble-mosaic courtyards that has become a victim of its own success – virtually every building is now a souvenir shop, bar, restaurant or guesthouse, and its beautiful crescent beach is crammed with sun-loungers and umbrellas. Slightly less crowded beaches can be found at **Lardos** and **Pefkos**, south of Lindos.

CHALKI ★

Chalki is tiny – 28km² (11 sq miles) – and most accommodation is at **Emborio**, its main harbour village, where the **Church of Agios Nikolaos** boasts the tallest bell tower in the Dodecanese. A ruined castle on a hilltop overlooks **Chorio**, the island's other village.

Right: *Tsambika is one of the best beaches on Rhodes, but very crowded in season.*

SYMI ★★

Treeless and barren, Symi made a good living from sponge-fishing, trading and shipbuilding until the Italian conquest of 1912 cut it off from the communities on the nearby Turkish mainland. Now its main village, **Gialos**, is half-deserted, with tumbledown tiers of gracious old mansions rising above a near-circular harbour. Gialos, and the **Monastery of the Taxiarchis** (St Michael) at **Panormitis** are popular with day-trippers from Rhodes. The best beach is at **Pedi**, a five-minute minibus ride from Gialos, and there are good restaurants here and at the harbour.

TILOS ★

The low hills of Tilos are crisscrossed by cobbled paths, making it a favourite walking island. Its main village, **Megalo Chorio**, is a charming muddle of whitewashed houses with a medieval fortress, a church dedicated to the Archangel Michael, and ruins of seven knightly castles. **Livadia**, the port, is the only other village.

Above: *Symi's tiered port is a favourite with day-trippers from Rhodes.*

NISSIROS ★

A bubbling cauldron of sulphurous mud lies in the centre of Nissiros, surrounded by the rugged walls of a dormant volcanic crater. On the crater rim, **Nikia** village is dazzlingly white and postcard-pretty, with great views. **Mandraki**, the harbour village, is famous for the **Panagia Spiliani** (Virgin of the Cave) chapel, within the walls of a tower built by the Knights of St John. On the hillside above, walls made of boulders mark the site of a Dorian settlement known as **Paliokastro**. There are beaches of black and brown sand at **Lies** and **Pachia Ammos**, and a few guesthouses and tavernas at **Mandraki**.

> **FOSSILS FOUND**
>
> Bones of deer, tortoises and dwarf elephant (*mastodon*) only 1.3m (4ft) tall, were found in 1971 in a ravine near the deserted village of **Mikro Chorio** on Nissiros. The island is believed to have been joined to Asia Minor until the Pleistocene period (time of the last ice age in Europe). More recent remains were revealed at **Agios Antonis** where there is a row of human skeletons on the beach thought to belong to villagers caught in lava in 600BC when Nissiros erupted.

KOS ★★

A long, thin, hilly island with long sandy beaches, a picturesque island capital, a quota of ancient and medieval sights and several lively resorts, Kos is one of Greece's

FATHER OF MODERN MEDICINE

Hippocrates (460–377BC), regarded as the father of modern medicine, was born on **Kos** and probably trained there. He travelled throughout the Greek world as a member of the **Asklepiada** (servants of Asklepios, god of healing) whose symbol – the staff (*caduesis*) – entwined by serpents, has survived. A vast collection of writings including the '**Hippocratic Oath**' is attributed to Hippocrates.

most popular mid-range package holiday destinations. The island's best beaches are at **Kardamena**, midway along the east coast, a former fishing village that has mushroomed into a full-grown strip of hotels, apartment complexes, bars, restaurants and tour agencies and is well known for its cheap and cheerful nightlife. **Kamari**, close to the island's southern tip, is a slightly more up-market and family-oriented resort on a sweep of sandy bay, below the tumbledown hill village of **Kefalos**.

The beaches on Kos's west coast can be windswept, and reliable *meltemi* winds rising on summer afternoons make it possibly the best windsurfing spot in the Aegean. The medium-sized resort at **Tingaki** is particularly well thought of by windsurfers and kite-boarders.

There are large, up-market hotel complexes either side of **Kos Town**, which has a busy fishing and commercial harbour overlooked by a fortress of the Knights of St John.

On **Platia Eleftherias**, a venerable, hollow plane tree is claimed to be the very one which sheltered the ancient healer and philosopher, **Hippocrates**.

Like Rhodes, Kos has a dwindling Muslim population, and the **Defterdar Mosque** (still used) and **Mosque of the Loggia** overlook Platia Eleftherias.

Along **Grigorou** (one block inland from the harbour) are the ruins of the Roman town, including a theatre, gymnasium, and 5th-century Christian basilica. On a hillside southwest of town, the theatre

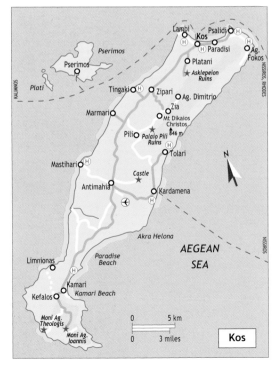

of the **Asklepeion** is well preserved. Relics from all of these sites can be seen in the **Archaeological Museum** on Platia Eleftherias (open 08:00–14:30 Tue–Sun).

KALIMNOS ★★

Rocky, barren and with a deeply indented coastline, Kalimnos has risen to tourism success on the coat-tails of its larger neighbour, Kos, but offers a much quieter holiday experience. Deep blue water surrounds its white limestone coasts, and there are good sand and pebble beaches at **Massouri** and **Myrties** on the west coast, at **Emborio** in the north and on the little island of **Telendos**, off the west coast. The main town, **Pothia**, is a colourful crescent of pastel-painted houses around a deep bay, formerly the base of the Aegean's biggest sponge-diving fleet – the island's divers travelled as far as Tunisia, Cuba and Florida in search of sponges, and Pothia is twinned with Tarpon Springs, in Florida, where there is an expatriate Kalymniot community.

LEROS ★

Wide, twin natural harbours cutting deep into its east and west coasts made Leros a fine naval base for Italy during its occupation of the Dodecanese and led to fierce fighting between Allied and Axis forces here in 1943. The crumbling Art Deco headquarters buildings (and a cinema built to entertain the troops) of the Italian fleet at **Lakki** are a fading colonial legacy. Courtesy of its airport, Leros gets a trickle of package tourism in the summer, and it's a good place for a relaxed holiday, with gentle hill-walking, a certain quirky charm, sandy beaches at **Panteli** and **Xirokambos**, some excellent old-fashioned fish tavernas and a medieval castle above the old port at **Agia Marina** that is still garrisoned by the Greek Army. Of all the Dodecanese islands, Leros really can claim to offer a slice of the 'real Greece'.

> **BEACHES ON KOS**
>
> **Psalidi** and **Agios Fokas** are heavily developed. **Embros Therma** has black sands, thermal springs and fewer visitors. **Kardamena** and **Mastichari** have good beaches but are crowded in season. The coast between **Kamari** and **Agios Stefanos** is dominated by a Club Med complex; both **Camel** and **Paradise** beaches are superb. The ruins of a 5th-century basilica of Agios Stefanos with mosaics and columns stand near **Kamari** beach.

Below: *Pothia, capital of Kalimnos, is a surprisingly large town with traditional Greek cafés and restaurants.*

Opposite: *Kastelorizo with its numerous islets forms Greece's eastern-most point.*
Below: *Ladies of Olympos in traditional dress keep alive the 'old ways' of island life.*

PATMOS ★★

The brooding fortress **Monastery of Agios Ioannis Theologos** (where St John the Divine wrote the Book of Revelations) casts its shadow over Patmos from its hilltop, drawing pilgrims from all over the Orthodox world. **Skala**, the island's harbour village, is a favourite with yacht sailors and has an attractive waterfront esplanade lined with smart café-bars. **Chora**, the hilltop village which surrounds the monastery, is a two-hour hike from Skala by a challenging long, steep cobbled path. Patmos is short of beaches, but 'water-taxis' go from Patmos to the pebbly bays of **Lambi**, on the north coast, and **Psili Ammos**, on a deep bay in the south.

LIPSI ★

Tiny Lipsi – 17km² (6 sq miles) – is the biggest of a mini-archipelago southeast of Patmos. With a 36km (25-mile) coastline, it has more than its fair share of beaches, and its only village stands beside a calm, glass-clear harbour full of colourful marine life. More than 50 tiny, blue-domed chapels dot its hillsides, and there are simple guesthouses and tavernas in its **Chora**.

ASTIPALEA ★★

Lonely Astipalea lies midway between the Dodecanese and the Cyclades and has more in common with the latter than with its eastern siblings. Its landscapes are barren and rocky, and **Chora**, its main village, is a Cycladic-style clutter of white buildings forming a maze of narrow lanes around a hilltop where an ancient acropolis, a Byzantine castle and a Venetian fort once stood. There are shingle beaches at **Livadia**, at the mouth of a lush valley, and at **Vathi**, the island's second port.

KARPATHOS ★★

Karpathos has been on the package holiday map for more than 20 years and receives direct charter flights, but seems oddly resistant to the influence of mass tourism. Its wild rugged landscape is quiet and unspoiled, and with mountain ridges rising to more than 1000m (3281ft) it is a paradise for walkers and naturalists. Ferries land at **Pigadia**, the main town and at **Diafani**, on the northeast coast. Until the early 1980s, Diafani provided the only access to the isolated hill village of **Olympos** – one of the last places in Greece where women still wear traditional costume on a daily basis – but a good, if vertiginous road now connects the mountain north with the rest of the island. **Ammopi**, between Pigadia and the airport, is a low-key package resort, as is **Arkasa** on the west coast. More small hotels and some authentically Greek restaurants are in Pigadia, where there is a long, shallow beach, and there are more beaches at **Kyra Panagia** and **Apella** at the foot of dramatic **Mt Kalilimni** (1215m; 3986ft), the highest mountain in the Dodecanese.

KASSOS ★

Karpathos's much smaller western neighbour is barren, depopulated, and bypassed by all but a handful of inquisitive visitors. **Fri**, a fishing village, is the island's capital. Kassos has no large beaches, but there is good swimming at **Chelathros** and great snorkelling around **Armathia**, a desert islet reached by caique from Fri. Stalactite-filled caves at **Sellai** and **Hellenokamara** are the island's other attractions.

KASTELORIZO (MEGISTI) ★

The remotest of all the islands, Kastelorizo lies less than 1km from the Turkish mainland. Badly damaged during a British assault during World War II, its only village is still mostly in ruins. Clear water (but no beaches) and the 'blue grotto' at Perasta are its main attractions.

The Dodecanese at a Glance

Despite spring sunshine, the sea is chilly till Jun and rain is possible. Jul–Aug are very hot and larger islands are crowded. Sep–Oct are usually sunny with cooler evenings and warm seas. Smaller islands are often crowded from late Jun to end Aug. With few attractions other than beaches, they are dull places when the sun is not shining; go in late May or Jun, or in Sep–Oct. Most hotels, guesthouses and restaurants in the Dodecanese are closed Nov–Apr.

Charter flights go to Kos and Rhodes from all over Europe from Apr–Oct, and to Leros and Karpathos from May–Sep. Olympic Airlines and Aegean Airlines fly to Rhodes and Kos several times daily from Athens; Olympic flies daily to Rhodes from Irakliou in Crete, and several times a week from Athens to Astipalea, Karpathos, Kassos, Kastelorizo, and Leros. Olympic also flies to Rhodes from Thessaloniki. **Ferries** and **hydrofoils** operate frequently between Rhodes and Kos, and less frequently from Kos via Patmos to Samos in the Northeast Aegean, to Mykonos, and via Astipalea and Amorgos to the western Cyclades and Piraeus. For up-to-date timetables consult www.gtp.gr Ferries operate at least daily from Piraeus to Patmos, Leros,

Kalimnos and Kos. Smaller ferries connect the Dodecanese 'satellite' islands (like Arki, Lipsi, Chalki and Symi) with their larger neighbours. On to the Northeastern Aegean, ferries run at least weekly between Rhodes, Kos and Samos, stopping at most points in between. **Ferry lines** include: **Blue Star Ferries**, tel: (Piraeus) 21089 19950; (Rhodes) 22410 76600; (Kos) 22420 28914; (Patmos) 22470 31324; (Leros) 22470 26000, www. bluestar ferries.com **DANE Lines**, tel: (Piraeus) 21045 29360; (Rhodes) 22410 43150. **Chalki-Kamiros Lines**, tel: 22410 45309. **Dodecanese Seaways**, tel: 22410 70590. **GA Ferries**, tel: 21045 82640.

Good, frequent **bus** services on Rhodes and Kos. Elsewhere they serve villagers and schoolchildren (first thing in the morning and late afternoon). **Taxis** are cheap and easy to find on all the islands except Lipsi, Kassos and Kastelorizo. **Cars**, **Motorbikes** and **scooters** can be hired on all but the smallest islands. Drive with caution.

On Rhodes advance booking is advisable in high season, especially for budget hotels which are in short supply. Lists of accommodation supplied by the Rhodes Tourism Promotion Organization, run by the island's hoteliers (see Useful

Contacts), and the Greek Travel Pages website, www.gtp.gr which also has online ferry and flight schedules. On smaller islands, accommodation in self-catering apartments, family-run guesthouses and small hotels is much more geared to the independent traveller; rooms are affordable and owners greet every arriving ferry. Booking ahead is advisable on Patmos all year round and essential during Greece's main religious holidays when the island fills with pilgrims. Almost all hotels (except for a few in Rhodes Town and Patmos) close from Nov–Apr. Top hotels include:

Rhodes

LUXURY

Rodos Park Suites, Riga Fereou 12, Rhodes New Town, tel: 22410 89700. Luxury suites with private balconies, choice of restaurants, large pool.
Miramare Wonderland, Ixia, Rhodes tel: 22410 96251, fax: 22410 95954. Secluded resort on one of Rhodes's finer beaches, not far from town.

MID-RANGE

Annapolis Inn, Oktobriou 28, Rhodes New Town, tel: 22410 31910, fax: 22410 31910. Excellent value, highly professional apartment-hotel in town.
Mansion Marco Polo, Agiou Fanariou 42, Rhodes Old Town, tel/fax: 22410 25562. Striking designer hotel with choice of huge rooms in main building or new self-catering apartments.

The Dodecanese at a Glance

Symi
MID-RANGE
Aliki, Gialos, tel: 22460 71665. Elegant hotel in a fine old ship-owner's mansion by the harbour; ask for sea-view room.
BUDGET
Les Catherinettes, tel: 22460 71671. Gialos. Delightful harbour view hotel in old neoclassical house. Very good value and very friendly.

Patmos
MID-RANGE
Patmos Paradise, Kambos, tel: 22470 32624, fax: 21042 24112. Village-style resort hotel; fine views, comfortable rooms.
BUDGET
Asteri, Skala, tel: 22470 32465, fax: 22470 31347. Friendly family hotel with garden near the pretty harbour.

Kos and Rhodes offer a huge choice of places to eat in main towns and resorts. For authentic Greek food, go where locals eat.

Rhodes: Rhodes Town
LUXURY
Alexis, Sokratous 18, tel: 22410 29347. Excellent fish restaurant, patronized by presidents and film stars.
Dinoris, Moussio 14A, tel: 22410 25824. Superb fish dishes in a medieval building.
MID-RANGE
Begleri, Zefyros beach, tel: 22410 33353. Excellent fish restaurant on the beach, with shellfish and sea urchins as well as more familiar seafood.

Lindos
MID-RANGE
Mavrikos, Platia, Lindos, tel: 22440 31232. Try lobster pasta or marinated whitebait at one of Rhodes's best restaurants.

Symi
LUXURY
Milopetra, Gialos, tel: 22460 72333. Classy restaurant; superb Mediterranean cooking, fresh fish and pasta.
MID-RANGE
To Ellinikon, Gialos, tel: 22460 72455. Imaginative menu combines eastern and western Mediterranean influences; excellent wine cellar.

Kos Town
MID-RANGE
Hamam, Nikita Nisyrou 3, Platia Diagoras, tel: 22420 28323. Interesting menu, picturesque old stone house.
Platanos, Platia Platanou, tel: 22420 28991. Modern Mediterranean cuisine next to Herodotus's aged plane tree.
Olympiada, Kleopatras 2, tel: 22420 23031. Classic, stick-to-the-ribs Greek cooking in this town-centre taverna.
Nikolas o Psaras, Averof 21, tel: 22420 23098. Kos's finest oúzeri (Greek equivalent of a tapas bar) serves oúzo, snacks.

Patmos
MID-RANGE
Aspri, Skala, tel: 22470 32240. A fine harbour view and great seafood.
Giagia, Skala, tel: 22470 33226. This Dutch-owned restaurant serves traditional Javanese-influenced dishes.

Rhodes: boat trips around the island and to Symi and Chalki; day trips to Marmaris in Turkey; scuba diving trips.
Kos: day trips to Bodrum in Turkey, Nissiros, Kalimnos and the small island of Pserimos.
From Patmos: boat trips to Lipsi and smaller islands.

Rhodes Airport Information, tel: 22410 88700.
Rhodes Tourism Promotion Organization, Ploutarchou Blessa 3, Rhodes Town, tel: 22410 74555-6, fax: 22410 74558, website: www.rodosisland.gr
Hellenic Tourism Organization (EOT), Rhodes, tel: 22410 23655 for information on all islands, website: www.justkos.com

DODECANESE	J	F	M	A	M	J	J	A	S	O	N	D
AVERAGE TEMP. °F	52	54	57	61	68	75	79	79	73	68	61	55
AVERAGE TEMP. °C	11	12	14	16	20	24	26	26	23	20	16	13
HOURS OF SUN DAILY	5	5	6	7	10	11	12	11	9	7	5	4
RAINFALL ins.	4.5	3	3	2	1	0.5	0	0	0.5	3	3	4.5
RAINFALL mm	114	73	70	43	18	8	1	0	13	73	82	110
DAYS OF RAINFALL	9	9	8	7	6	3	1	2	2	6	9	9

7
The Northeast Aegean

The islands of the northeast are the Aegean's final tourism frontier. Bigger, and further apart, than their little cousins in the Cyclades, each has a clear, individual identity.

Samos, with its verdant hills, vineyards and pebble coves, and **Lesbos**, with its big landscapes, sandy beaches and pretty villages, are the high-profile holiday isles, along with pine-covered **Thasos**, close to the northern mainland. **Ikaria** has fantastic beaches, great walking, and a loyal summer-bohemian following. **Chios** and **Limnos** are only beginning to open up to tourism, despite being easily accessible by air and sea. And **Samothraki**, with its eerie ancient temple site and the highest mountain summit in the Aegean, remains resolutely off the beaten track – as do the tinier satellite islands of the Northeast Aegean, including **Psara**, **Inousses** and lonely **Agios Efstratios**. Even **Fourni**, with frequent ferries from Piraeus and from its bigger neighbours, is only beginning to be discovered.

The northeast remains well off the beaten islandhopper's trail – mainly because inter-island ferries are infrequent and crossings are long. That said, each of these big, and relatively unknown islands repays exploration, whether on foot, by bicycle, rented car or island bus, and there is plenty to occupy the visitor for a week or a fortnight on Samos, Chios, Lesbos or Limnos, while the smaller islands are perfect for a totally relaxed holiday. Thasos, with some of the best white, sandy beaches in the Aegean, is ideal for families with small children.

ALBANIA
TURKEY
●ATHENS
MEDITERRANEAN SEA

DON'T MISS

★★★ **Chios:** unique Pirgi and the houses of the other mastic villages with their lovely geometric painted walls.
★★★ **Samothraki:** sanctuary of the Great Gods – one of the most impressive yet least-visited ancient sites in the Aegean.
★★ **Lesbos:** medieval hill village of Agiassos and Mt Olympos – mountain walks with spectacular views.
★ **Thasos:** an island ringed with lovely beaches.

Opposite: *The geometrically decorated houses of Pirgi, one of the mastic villages on Chios.*

Above: *Close to the shore at Pithagorio stands a castle, built by Logothetis, a hero of the 19th-century revolution against Turkish rule.*

SAMOS ★★★

The lushly wooded hill-sides of Samos's central massif, **Mt Ampelos** (1140m; 3740 ft), over-looking coastal plains covered with olive groves and vineyards that pro-duce some of Greece's best white wines, are in sharp contrast to the bare stone summit of **Mt Kerkis** (1445m; 4741 ft) which looms over the west end of the island. With its cypress trees, tidy farms, and villages of elegant, red-roofed houses and Italianate campaniles, Samos reminds some visitors of Tuscany.

This large island (472km²; 182 sq miles) appeals to everyone, with an array of ancient and medieval relics, long beaches, clear blue water, charming resorts and villages as well as larger towns that offer a slice of Greek life that is not beholden to tourism.

Pithagorio, looking across the strait to nearby Turkey, is the island's busiest resort town, with a yacht marina and a bustling fishing harbour and ferry port lined with tall, pastel-painted buldings. Named after the mathematician Pythagoras, this was the site of the island's capital in classical times and archaeologists have uncovered lengths of the ancient walls, traces of Roman-era baths

---

PYTHAGORAS

Pythagoras is universally known for the mathematical theorem bearing his name. He was born some time around 580BC and relied on his followers to document his teachings. His ideas on pro-portions in beauty influenced Classical sculpture and architecture. His thoughts also influenced the develop-ment of mathematics and its application to music and astronomy and he believed in the transmigration of souls.

and a theatre, and the 1km (900yd) **Efplinion tunnel** that carried water to the city. From the theatre a road leads to the **Church of Panagia Spiliani** (the Virgin of the Cave) in a grotto which in pre-Christian times was the cave of an order of Sybils (prophetesses). Overlooking the town is the castle of **Lykourgos Logothetis**, who led Samian forces during the War of Independence (though Samos, like its Northeast Aegean siblings, remained under Turkish rule until 1913).

West of Pithagorio, a solitary column and rows of foundation stones mark the site of the **Temple of Hera** (open 08:00–14:30, Tue–Sat), an ambitious project which was begun in 525BC but never completed. Beside it, the tiny fishing hamlet of **Irion** has ballooned into a small resort beside a long pebble beach. To the east of a range of hills dotted with small farming villages, the little port of **Ormos Marathokampos** has also blossomed into a low-key resort, and a ribbon of small hotels and holiday tavernas stretches along a huge sweep of sand and shingle from **Votsalakia** and **Psili Ammos** to **Limnionas** and the foot of **Mount Kerkis**, which offers demanding hiking – only experienced and well-equipped mountain walkers should try for the summit.

On the north coast, the island capital, **Vathi**, lies on a deep inlet and is a bustling commercial centre with some dignified old buildings along its waterfront and its main square. Finds from Pithagorio and the Temple of Hera are displayed in the good, modern **archaeological museum** (open 08:00–14:30, Tue–Sat). East of Vathi, the **Monastery of Zoodochos Pygi** stands on a headland above the fishing village of **Mourtia** – from here on clear days Turkey seems almost close enough touch.

Below: *Fertile Samos seems more lush than other Aegean Islands and the land is filled with olives, cypress trees and pines.*

DAEDALUS AND IKARUS

Daedalus was the ingenious inventor and engineer who designed the labyrinth for King Minos of Crete. Unwisely, Daedalus aided the king's daughter Ariadne in saving Theseus. Minos ordered his men to watch all ports and so, in order to escape, Daedalus constructed wings for himself and his son Ikarus using feathers and wax. They flew from Crete but Ikarus flew too close to the sun, the wax melted and he plunged into the sea. The island of Ikaria rose where he fell.

Below: *With limited development of tourism, quiet places are easy to find in Ikarian villages.*

Midway along the north coast, **Kokkari** is the island's most picturesque resort, with good hotels, tavernas and cafés spread out along a pretty waterfront between two headlands, and white-pebble beaches nearby at **Lemonakia** and **Tsamadou**. **Agios Konstandinos**, a few kilometres west, has also grown into a low-key holiday resort with a long, pebbly beach.

Karlovasi, at the western end of the island, is a spread-out market town which is also home to several faculties of the University of the Northeast Aegean. There is a long, but untidy pebble beach at **Limani**, Karlovasi's commercial harbour, which despite drab and unkempt air has several excellent fish tavernas which are favoured by locals.

IKARIA ★★

Long, thin and mountainous, Ikaria has some of the Northeast Aegean's longest, finest and least crowded beaches near **Armenistis** (reached from the north coast port of **Evdilos**) and is an excellent island for walking, with green valleys and a high moorland plateau watered by small streams above the orchards and vineyards of the pretty mountain village of **Christos Rahes**. The spa at **Therma**, next to the sleepy south coast capital, **Agios Kirikos**, is highly radio-active but favoured by elderly Greeks.

FOURNI ★

Between Samos and Ikaria, little Fourni's magnificent, secret harbour was a corsair's lair until the 18th century. Today it is the home port of the Aegean's largest fishing fleet, with more than 100 wooden caiques sending their catch to the restaurants of Athens. Most islanders live in the harbour village,

Kampos, from where it is a stiff walk over steep hillsides to sandy coves at **Psili Ammos**, **Elidaki** and **Kambi**.

CHIOS ★★

Chios is the ancestral home of some of Greece's wealthiest maritime dynasties, and its aloof attitude to tourism is attributable to its fertile farmland and mercantile wealth. Coastal plains surround a mountainous interior of pine forests and barren slopes scarred by forest fires, and the southern part of the island is dotted with the picturesque settlements known as *mastiko-choria* (mastic villages). Chios has clean, pebbly beaches at **Lefkada**, **Managros** and **Elinda** on its northwest coast, and at **Emborios** on the south coast, but has no real beach resorts.

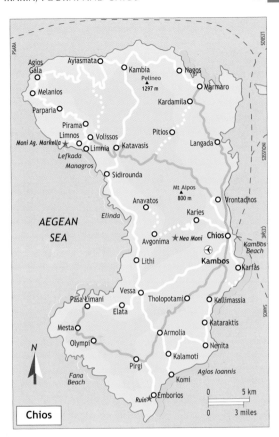

Chios

Chios, the capital, is a no-nonsense port town built around a busy working harbour which is overlooked by the dilapidated battlements of a Byzantine-Genoese fortress. Genoese merchant princes snapped up the island in 1346, and their legacy survives in the **Philip Argenti Museum**, founded in 1932 by a scion of one of the last Genoese-Chiot dynasties. Within the walls of the Genoese Kastro, the **Giustiniani Museum** (open 09:00–15:00 Tue–Sun) is named for another great Aegean-Genoese family and has a fine collection of early

MASTIKOCHORIA

The climate of southern **Chios** has long suited the growth of mastic trees (*Pistacia lentiscus*). An aromatic resin obtained by bleeding the bark from the evergreen tree, mastic was highly prized and sought after since ancient times as an astringent, a general cure for most ailments and as an ingredient in perfumes, chewing gum and sweets. The fragrant wood was used for toothpicks. Mastic is still used in lacquers and varnishes, although demand for mastic gum has dropped since the arrival of synthetics. From *mastikhan* – to grind the teeth.

Above right: *Nea Moni, home of a revered icon.*
Below: *The decorated tradition (xista) of the mastic villages was taught to the locals by the Genoese.*

Christian mosaics, frescoes and icons. The cemetery of the former Mecidiye mosque (now a storehouse for antiquities awaiting restoration) with its Armenian, Turkish and Jewish gravestones, bears testament to a vanished, multi-ethnic Aegean world.

South of Chios is the **Kambos**, a miniature fertile plain covered in citrus groves and crisscrossed by stone walls and sunken lanes which divide the former estates of the Chiot aristocracy. Inland, **Nea Moni** (New Monastery) is in fact almost 1000 years old; it was built by the Byzantine Emperor Constantine VIII Monomachos on the site of an earlier foundation where a miraculous icon of the Virgin was found. Its 11th-century mosaics were reassembled after an earthquake in 1881 which collapsed the dome. The deserted medieval village of **Anavatos** is reputedly haunted by the ghosts of villagers who threw themselves from the cliffs below rather than fall into Ottoman hands during the massacres of 1822, during the War of Independence, in which 30,000 Chiots died and 60,000 were enslaved.

The island's most outstanding areas of interest, however, are the 'mastic villages' of the south. Built inland to protect the villagers from marauding corsairs, each village was ringed by watchtowers and the outer ring of houses acted as protecting wall. Within, a maze of narrow lanes was designed to baffle raiders. **Mesta** is the best preserved of the *mastikochoria*. Even more photogenic is **Pirgi**, with its striking and unique xysta wall paintings in geometric patterns of black and white.

PSARA ★

It is hard to believe that desolate Psara, around 32km (20 miles) west of Chios, was once a wealthy island with a population of more than 30,000. During the War of Independence it was briefly the capital of the Greek revolutionary government, until in 1824 the Turks assaulted the island, killing or enslaving most of its people and driving the survivors into exile. Only a handful of people live here, and although it has fine beaches it sees very few visitors – making it the best spot in the Northeast Aegean for solitude seekers.

LESBOS (MITILINI) ★★★

Lesbos is the most appealing of all the Northeast Aegean isles, with big landscapes, fine beaches, medieval castles

Left: *Anaxos village lies on a sweeping bay – peace and quiet a short distance from Molyvos at the northern tip.*

EVANGELISMOS

The convent of Evangelismos on Inousses was built in the 1960s in memory of Irini, 20-year-old daughter of Katingo Pateras, of a prominent ship-owning family. The girl died from Hodgkinson's disease after praying to take the illness from her father who was suffering from it. After three years when her body was exhumed (local custom) it was found to be mummified, convincing her mother that she was a saint.

Below: *The donkey is still the traditional beast of burden used to carry the fruit and olive harvests.*

and pretty villages. The third largest of the Greek islands (1630km²; 629 sq miles), it caters to all tastes, from the families with toddlers who favour the warm, shallow waters and gentle sandy beaches of the **Gulf of Kalloni** to the lesbian women who make the summer pilgrimage to **Skala Eressou**, birthplace of the poetess Sappho. Rare flowers, migrant birds, and a network of old hill tracks attract walkers, naturalists and painters.

Molyvos (Mithymna), midway along the north coast, is Lesbos's showcase village, with streets of red-tiled stone houses spilling down the slopes below a spectacular medieval **Genoese castle** (floodlit at night, open 08:30–15:00, Tue–Sun) to a small harbour and a pebble beach. A longer, sandier beach at **Petra**, 7km (4.5 miles) south-west of Molyvos, has turned this former farming village into the nucleus of a low-key resort. To the east, **Skala Sikamias** is a pretty fishing village with a handful of places to stay and good fish restaurants, along with natural hot springs rising on a pebble beach.

Beaches on the north coast beyond Petra are windy and unpleasantly littered with trash that floats across from Turkey, but **Sigri**, on the west coast, is a pleasant small resort with a sandy beach and a harbour guarded by an old Turkish fort. South of Sigri, the 'petrified forest' of fossil tree trunks scattered around the hillside and beach is one of the island's less exciting attractions. **Skala Eressou**, further south, has a small fishing port with a choice of tavernas and small guesthouses, and a mile-long pebble and sand beach. The **Gulf of Kalloni**, an almost landlocked bay, cuts deep into the west coast. Shoals of sardines thrive in its warm, shallow waters, and its calm, gently shelving beaches are ideal for families with toddlers. Saline marshes and saltpans at the inner end of the gulf attract migrant wading birds including black-winged stilts, avocets, black and white storks and the occasional pelican.

Inland, pines and chestnuts cloak the slopes of **Mt Olympos** (968m; 3176ft) which dominates the south of the island. On its eastern side, the old-fashioned village of **Agiassos** is crowned by a medieval castle and surrounded by plum orchards and walnut groves, and there is a fine, long walk down to the coast at **Plomari** via **Paleokastro** and **Megalochori**. There are more walks around **Mt Lepetimnos** (968m; 3176 ft).

Mitilini, the island's capital, is a commercial port with a scattering of Roman, Byzantine and medieval relics, including a Byzantine-Genoese fortress and the remains of a Roman aqueduct, ancient city walls, and a Hellenistic theatre. Finds from archaeological sites around Lesbos, including Roman reliefs and Greek mosaics, are displayed in the **Archaeological Museum** (Open 08:00–15:00 Tue–Sun).

Above: *A spectacular view of Mirina, port and capital of Limnos. It is often called Kastro for its castle built over the promontory in the midst of the sandy shore.*

LIMNOS ★

A tiny yet picturesque Genoese castle stands on a crag dominating the harbour at **Mirina**, capital of this big, low-lying island of wheat and tobacco fields. **Ormos Moudras** (Moudros Bay) is the finest deep natural anchorage in the Aegean and is a major base of the Greek Navy, and a large military presence domi-

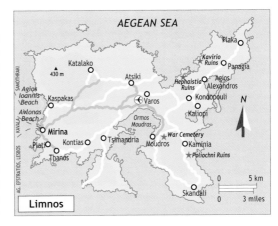

nates the island. A long, sandy beach with a few hotels stretches from the castle promontory northward to **Akti Mytrina**; the rest of the island is untouched by tourism. **Mirina** is a pretty, sleepy town with a waterfront linedby elegant old Ottoman buildings and some excellent fish taverns. Its recently renovated **Archaeological Museum** (open 08:00–14:00, Tue–Sun) is surprisingly interesting, with fine displays of finds from Limnos's archaeological sites. These include **Hephaistia**, the largest city on the island in Hellenistic times; **Kavirio**, a sanctuary to the enigmatic great gods; and the 4100-year-old fortified settlement at **Poliochni**.

Above: *Doric columns re-erected at the Sanctuary of the Gods at Samothraki which marks the Hieron where final stages of initiation took place.*

AGIOS EFSTRATIOS (AY' STRATIS) ★

Agios Efstratios (St Eustace) is said to have died on this loneliest of Aegean isles, midway between Limnos and the Sporades, and there are reportedly moves afoot to twin the island with Saint Eustatius, in the Dutch Antilles. Only 43km² (17 sq miles) in size, Agios Efstratios is as far off the tourist track as it is possible to get; its harbour is too small for large vessels, and passengers from the regular Rafina-Limnos ferry are disembarked by smaller boats.

SAMOTHRAKI ★

Green and mountainous, Samothraki is a well-kept secret, seeing few visitors other than a handful of summer cognoscenti and a trickle of scholars bent on viewing the remarkable **Sanctuary of the Great Gods**, a pre-Olympian group of deities who were worshipped here centuries before Greeks from Samos colonized the island around 700BC. Almost circular, the island rises steeply to the 1611m (5236 ft) summit of romantically named **Fengari** ('Moon Mountain'), from which the god Poseidon, according to legend, watched the progress of the Greek siege of Troy.

Fengari is the highest peak in the Aegean except for the mountains of Crete, and is sometimes snow-capped as late as April. **Kamariotissa**, the port, is functional and comes to life only to load and unload ferries and freighters. **Chora**, the main village, is crowned by the remains of a Byzantine tower, with red-roofed stone houses and narrow paved streets. Bohemian summer visitors gather at **Therma**, at the foot of Fengari, where there are a few small guest-houses, camp sites beneath chestnuts on a long pebble beach, mountain-walking trails and hot springs which attract elderly Greeks seeking relief from their aches and pains.

THASOS ★★★

Beach lovers are spoilt for choice on Thasos. This near-circular, pine-wooded island within sight of the northern mainland is ringed by beaches of deep, soft white sand that make it a family favourite, among them **Psili Ammos** ('Silver Bay'), **Astris**, **Potos**, and **Pefkari**.

A Hellenistic theatre on the slopes of the ancient Acropolis overlooks the main harbour town, **Limenas**, and other ancient relics here include the remnants of the Roman Agora, Odeon, and Sanctuary of Hercules. On the Acropolis are the tumbledown walls of a Genoese fortress, foundations of a Temple of Athena Paliouchos, and a steep 'secret stair' leading to a Sanctuary of Pan, the goat-footed god of wild places.

Limenaria, Thasos's second largest town and its main resort area is very popular in summer because of its long stretch of well-shaded, sandy beach, clear blue water and plentiful water sports.

> ### BLACK WINE
>
> Thasos was reputed to be the birthplace of the goddess of agriculture, Demeter. She favoured the island and caused its grapes to produce the famous 'black wine'. Wily Odysseus used a supply of this wine to overcome the one-eyed Cyclops – the giant who had captured the hero and his crew.

Below: *Looming mountains form a dramatic backdrop to the numerous sandy beaches of Thasos.*

98

The Northeast Aegean at a Glance

BEST TIMES TO VISIT

Spring arrives later in the
Northeast Aegean isles than in
their southern cousins and it
can be cold, wet and and
windy as late as May. Late Sep
and Oct are cooler than in
Crete or Rhodes and winters
are cold – snow can fall on the
mountains of Ikaria, Samos
and Samothraki. Samos is very
crowded in Jul–Aug. Other
islands have a much shorter
holiday season and are busy
only from Jul to late Aug.

GETTING THERE

Charter flights bring package
tourists from many European
cities to Samos and Lesbos
May–Sep; charters also fly into
Kavala, on the northern main-
land, for Thasos (a 30-minute
ferry transfer). Olympic Airlines
flies daily from Athens to Chios,
Ikaria, Samos, Lesbos, and
Limnos, and at weekends from
Thessaloniki to Chios, Samos,
Limnos and Lesbos. There are
some direct inter-island flights
between Chios, Limnos,
Mitilini and Samos, but beware
flights requiring a transfer in
Athens. **Ferries** operate at least
four times weekly (mostly
overnight) from Rafina and/or
Piraeus to Ikaria, Fourni,
Samos, Chios, Lesbos, and
Limnos (calling at Agios
Efstratios). Shuttle ferries
operate all day between Kavala
and Thasos. Ferries go several
times weekly from Kavala or
Alexandroupolis on the northern mainland to Samothraki

and on through the Aegean
islands to Samos. Samos is the
connection for ferries and (in
summer) hydrofoils southward
to Patmos, Kos and the rest of
the Dodecanese, and west to
Mykonos and the Cyclades.
Ferry lines include: Chios-
Oinousses Lines, tel: 22710
41390. GA Ferries, tel: 21045
11720 (all the main northeast
Aegean islands). Hellenic
Seaways, tel: 21041 99000
(Samos-Fourni-Ikaria-Patmos).
Kavala Shipping, tel: 25108
31130 (Kavala-Thasos).
Kaveiros Lines, tel: 25510
26721 (Alexandroupoli-
Samothraki). Kiriacoulis Lines,
tel: 22730 3232 (Samos-
Patmos-Kos). Miniotis Lines,
tel: 22710 24670 (Chios to
Ikaria, Samos, Psara, Turkey).
NEL Lines, tel: 22510 26299
(Lesbos to Piraeus, Limnos,
Samothraki, Alexandroupoli,
Kavala, Thessaloniki).
Samothraki Lines, tel: 25510
26721 (Samothraki to
Alexandroupoli and Kavala).
Sunrise Lines, tel: 22710 41390
(Chios-Cesme). ANETH-
Thassos Shipping, tel: 25930
22318 (Thasos-Kavala).

GETTING AROUND

Good **bus** and **taxi** services on
all these large islands; **car**,
scooter and **motorcycle** hire
available in all resorts.

WHERE TO STAY

Rooms can be hard to find on
Samos, Lesbos and Thasos in
high season – these islands are

geared mainly to package
tourism and most hotel rooms
and apartments are block-
booked for the entire season.
Smaller islands such as Psara,
Fourni and Ikaria have limited
accommodation but guest-
house owners meet all summer
ferries to offer beds. See www.
gtp.gr for up-to-date listings.
Recommended places include:

Samos
LUXURY
Arion, Kokkari, tel: 22730
92020, fax: 22730 92006.
Exclusive village-style resort set
in gardens above the village.

MID-RANGE
Samos Hotel, Them.Sofouli
11, Vathi, tel: 22730 28377,
fax: 22730 28482. On harbour,
ideal if arriving late or leaving
early by ferry. Rooftop pool.

Ikaria
MID-RANGE
Cavos Bay Hotel, Armenistis,
tel: 22750 71381, fax: 22750
71380. Medium-sized hotel
with all mod cons including
pool, great views.

Chios
LUXURY
Perleas Mansion, Vitadou,
Kambos, tel: 22710 32217,
fax: 22710 32364. Gorgeous
converted 17th-century manor
with pool, set in a huge estate.

Lesbos
LUXURY
Pyrgos of Mytilene, Venizelou

The Northeast Aegean at a Glance

49, Mitilini, tel: 22510 27977. Opulent, VIP hotel, in centre of Lesbos's main port – for a pampered overnight stay but not for a fortnight's holiday.

Limnos
LUXURY
Akti Myrina Hotel, Myrina Beach, tel: 22540 22681, fax: 22540 22947. De luxe, village-style resort on a fine beach – the best resort hotel in the Northeast Aegean.

WHERE TO EAT

Fertile farmland and rich fishing grounds mean the Northeast Aegean isles offer some of the best fresh local produce in Greece. Fourni is a seafood-lover's paradise, with fresh fish and shellfish in a row of friendly, simple harbour-front tavernas. The harbours of Chios and Limnos also have a great choice of fish tavernas. The resorts of Samos and Lesbos serve the usual tourist fare, but there are authentic tavernas and grills in the less-touristed port towns of Vathi and Karlovasi on Samos and Mitilini on Lesbos. Outstanding places to eat include:

Samos
Riva, Pithagorio, tel: 22730 62395. Imaginative, authentic Greek cuisine served in an old house next to the castle.
Pigi Pnaka, Vourliotes, tel: 22730 93380. Fabulous traditional village tavern under spreading plane trees.

Lesbos
Chryssi Akti, Molyvos, tel: 22530 71879. Good taverna on Molyvos's pebbly Eftalou beach.
Ermis, Kornarou 2/Ermou, Mitilini, tel: 22510 26232. May be the best old-fashioned oúzeri in the islands; fabulous snacks and a great place to kill time.

Chios
O Moriasis ta Mesta, Mesta, tel: 22710 76400. The finest taverna in the medieval village; Greek food (rabbit stew, salads of wild herbs) with a difference.

Limnos
Mouragio, Myrina, tel: 22540 41065. Fishermen's tavern that has been smartened up; great seafood, right on the harbour.

TOURS AND EXCURSIONS

Samos: picnic and snorkelling trips to the uninhabited island of Samiopoula, off the south coast, from Pithagorio and Ormos Marathokampos. Cruises to Fourni from Ormos and Karlovasi. Day trips by hydrofoil in summer to Patmos from Pithagorio; day trips to Kusadasi in Turkey with excursions to ancient Ephesus, also from Pithagorio.

Chios: day trips to Cetme in Turkey, picnic and snorkelling cruises to Inousses, a desert isle a few kilometres east of Chios, reputedly the wealthiest spot in the Aegean and home to a reclusive clan of wealthy ship-owning familes, one of which, the Pateras, endowed the island's Evangelismos convent.

Thasos: cruises around the Athos peninsula, Greece's 'Holy Mountain'; coach tours from Kavala on the mainland to the Hellenistic ruins of Philippi.

USEFUL CONTACTS

North Aegean Islands Tourism Directorate, Aristarchou 6, Mitilini, Lesbos, tel: 22510 42511. **Samos Tourist Office**, Vathi, tel: 22730 28582. **Chios Tourist Office**, Chios, tel: 22710 44389. Lesbos Tourist Office, Mitilini, tel: 22510 42511 **Lesbos Tourist Office**, Molyvos, tel 22530 71347.
Chios: www.chios.com
www.chioscity.gr
Lesbos: www.e-lesbos.com
www.molivos.net
www.about-lesbos.com
Samos: www.samoshotels.com
Thasos: www.thassos-island.gr

AEGEAN ISLANDS	J	F	M	A	M	J	J	A	S	O	N	D
AVERAGE TEMP. °F	43	45	50	57	66	73	79	77	70	61	52	46
AVERAGE TEMP. °C	6	7	10	14	19	23	26	25	21	16	11	8
HOURS OF SUN DAILY	4	5	6	7	10	11	12	11	9	7	5	4
RAINFALL ins.	3	2.5	2	2	2	1	1	0.5	1	3	4	4
RAINFALL mm	71	63	47	48	40	26	19	15	28	73	95	93
DAYS OF RAINFALL	5	5	5	6	5	4	2	2	2	5	8	7

8
The Sporades

A total of five islands are grouped together for administrative purposes to make up the Sporades or, in Greek, the 'scattered islands' – a name that in ancient times comprised all the islands situated beyond the circumference of the Cyclades. Three of these – **Skiathos**, **Skopelos** and **Alonissos** – lie close together, not far from the Pelion Peninsula of mainland Greece. A fourth, **Skyros**, sits aloof in the Aegean and has little in common with these or any other island group. **Evia** – the second largest Greek island and the fifth in the pack – is actually more like an extension of the mainland (*see* p. 37).

Skiathos, with an international airport, charter flights from the UK and elsewhere in Europe, and some of Greece's best sandy beaches, is the gateway to the **Sporades** and the most popular island in the group. Its neighbours, **Skopelos** and **Alonissos**, are connected to Skiathos by frequent summer hydrofoils and ferries. Although they lack Skiathos's sandy beaches, they attract the kind of audience seeking peace, quiet, and lovely scenery.

The Sporades are not the kind of islands popular among typical sightseers – they have no landmark ancient or medieval sites, and only a scattering of old villages and a few Orthodox shrines and chapels. That said, for those with itchy feet, the mainland port of **Volos** is conveniently close (it can be reached in just 90 minutes by hydrofoil from Skiathos) opening the way to the northern mainland.

DON'T MISS

★★★ Alonissos: the islets and the chance of seeing monk seals which breed in Piperi.
★★★ Skiathos: wonderful beaches, overwhelmingly popular in summer.
★★ Skopelos: its island walks always in sight of the sea.
★★ Skyros: the Faltaits Museum with its intriguing collections of folk items: embroidery, porcelain and carved wood.

Opposite: *An unspoiled island, the beauty of Skopelos is exemplified by Stafilos Beach.*

ISLAND TRADITIONS

A traditional example of a Skyriot house can be found in the **Archaeology Museum**. The house is an incredible feat of compact living to rival a Romany caravan with space of around 35m² (377 sq ft), decorated with beautifully carved furniture, ornate lattice work and inumerable ornamental plates. Since the 16th century plate collecting has been the national Skyriot passion. Open Tuesday–Sunday 08:00–14:30.

The **Faltaits Museum of Folklore** is located in the northeast of Skyros, high on a ridge. It has a marvellous collection of bits and pieces amassed by Manos Faltaits, painter, poet and writer. Its shop has excellent reproductions for sale. Open daily 10:00–13:00; 17:30–20:00.

SKIATHOS ★★★

Idyllic pine-fringed beaches attract up-market package operators to the island of Skiathos – out of season the wooded 61km² (24-sq-mile) island is extraordinarily beautiful and friendly.

There is only one main settlement here – **Skiathos town** with its picturesque red roofs and sparkling, white buildings. The town sits in a bay strewn with islets and its twin harbours have numerous waterfront restaurants including one in the Venetian Kastro on the Bourtzi promontory.

Jam-packed buses ply their trade on the south coast road – with stops for **Megali Ammos**, **Achladies**, **Kalamaki** and **Kanapitsa**. The north coast beaches tend to be less crowded since they are affected by the *meltemi*. **Lalaria**, reached by sea, has a natural arch at one end and several sea caves.

Paths leading across the island offer an escape from the beach to **Mt Karafiltzanaka** (411m; 1349ft), the highest point of the island, and to the peaceful **Evangelistria Monastery** (open daily from 08:00–12:00 and 17:00–19:00).

Right: *Many of the beaches on Skiathos are pine-fringed. Shown here is Koukounaries and its port.*

SKOPELOS ★★★

Tourism has not grown as rapidly on **Skopelos** as it has on some of its neighbours. This is a lovely, relaxed 'family' island (96km²; 37 sq miles), with extensive pine forests and pebble beaches.

Skopelos town (or Chora), the main settlement on the island, is situated on the exposed northern coast and hydrofoils and ferries make for a small quay at **Agnondas** on the southern side in rough weather. The town itself suffered some damage in the 1965 earthquake

Above: *Red-roofed Skopelos town forms the main settlement and harbour for the island.*

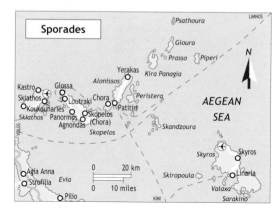

Right: *Picturesque Patitiri, where the Hellenic Society for the Protection of the Monk Seal has an office.*

but has nevertheless retained its character. Within the town there are the remains of a Venetian Kastro and numerous churches – **Zoodochos Pigi** has an icon attributed to St Luke.

The town is known throughout the region for its prunes – plums are dried in a large oven at **Fournou Damaskinon**. Traditionally crystallized prunes are served with *raki* (see p. 27).

Regular buses connect Skopelos with the coastal settlements – **Stafilos** has an ancient Minoan tomb, **Panormos** is a favourite with windsurfers; **Milia** and **Klimna** tend to be quieter. Other beaches on the island can be reached in the high summer season by caique – **Glisteri** from Skopelos and **Limonari** from Agnondas. **Glossa**, which is famed for its almonds, is a pretty village situated just above the small port of **Loutraki**.

ALONISSOS ★★

The first national marine park in Greece was established around **Alonissos** (62km²; 24 sq miles) and the nine islets in its archipelago, in order to offer protection to the monk seal.

A severe earthquake decimated the main town in 1965. Citrus production also fell sharply and recovery has been very slow. **Alonissos town** (Chora) has a magnificent setting, with its Byzantine walls and memorable sunsets.

The best beaches on the island can be reached by caique: **Chrysi Milia** has a sandy pine-backed beach, while **Kokkino Kastro** is composed of shingle and is also the site of the former capital of ancient **Ikos** (an early name for the island). Some beautiful unspoiled beaches are to be found and explored at **Yerakas**, **Kopelousako** and **Megaliamos**.

For fishing and water sports try **Steni Vala** and **Kalamakia**; snorkellers can find traces of an ancient settlement in the sea near **Agios Dimitrios** and a sunken Byzantine boat at **Agios Petros**.

SKYROS ★★

Most of the inhabitants of this island live in the central portion of the island, particularly around the port of **Linaria** and the capital, **Skyros**. To the north the land is well forested but to the south it is dry and barren with steep cliffs.

Skyros (Chora) is Cycladic in appearance and its houses encircle an acropolis surmounted by a kastro. Poet Rupert Brooke is buried at **Tris Boukes** on the southern tip of the island and **Brooke Square** in Skyros town is named in honour of him.

Sandy **Magazia** and **Molos** beaches, located near the town, are the busiest: **Papa ta Chomata** and **Ormos Achilli** are also within walking distance. The best beaches, however, are in the north at **Agios Petros**, **Pefkos**, **Acherouses** and **Agios Fokas**. **Skyropoula**, with its two beaches and cave, is reached by caique from **Linaria**.

THE ARCHIPELAGO

The other islets in the archipelago can be visited by caique from Patitiri but generally only in the high summer season:
- **Peristera:** lovely, sandy beaches and olive groves.
- **Pelagos (Kira Panagia):** abandoned monasteries and thick pine woods.
- **Psathoura:** home of the Sirens, the mythical songstresses who enchanted Odysseus. The island has a tall lighthouse and the remains of an ancient city beneath the sea.
- **Skantzoura:** sea caves, coves and excellent fishing.
- **Yioura:** an ancient breed of goats and the Cyclops cave (from mythology).

The Sporades at a Glance

In the Sporades, **summer** temperatures are softened by cooling sea breezes and the skies are generally clear. **Winters** are mild but enough rain falls to support the extensive pine forests and rather lush vegetation.

Charter flights can be taken in summer to Skiathos from the UK and also from some mainland European airports. Olympic Airlines flies at least once a day from Athens to Skiathos, and also twice weekly to Skyros.
Minoan Lines (Skiathos, tel: 24270 22018; Athens, tel: 21092 70150; Agios Constantinos, tel: 22350 33376) runs a luxury **coach** service from Kalimarmaro Stadium, Vas. Constantinou 2, central Athens, to connect with ferry or hydrofoil services to Skiathos and Skopelos from Agios Constantinos on the mainland. **Ferries** from Rafina and **hydrofoils** from Volos go (several times daily in the summer months) to Skiathos, Skopelos, Alonissos and at least daily between Kymi (Evia) and Skyros. A weekly service from Thessaloniki calls at all four

islands and onward to the Cyclades. There are also smaller ferries that ply the route from Skiathos to Agios Constantinos on the mainland. Note that Skopelos has two ports: there is one at Chora, the main town, and another at the tiny village of Glossa at the opposite end of the island. Boats from Skiathos call at both these harbours. Ferry lines include:
GA Ferries, tel: 21045 82640 (Thessaloniki-Sporades-Cyclades).
Hellenic Seaways, tel: 21041 99000, www.hellenic seaways.gr (Volos-Skiathos-Skopelos-Alonissos).
Skyros Shipping, tel: 22220 91789 (Kymi-Skyros).
Port Authorities:
Skiathos, tel: 24270 21885.
Alonissos, tel: 22424 65595.
Skopelos, tel: 24240 22180.
Skyros, tel: 22220 91475.

Buses are infrequent on all four of the Saronic islands – Skiathos, Skopelos, Alonissos and Skyros – but **taxis** are good value for money, distances are short, and **motorcycles**, **cars** and **scooters** are readily available and can easily be hired on Skiathos, Skopelos and Skyros.

Accommodation is somewhat limited on all four of the islands and booking in advance is advisable. Most of the hotel accommodation on Skiathos is usually block-booked by package holiday companies, which means that accommodation is sometimes difficult to find. Skopelos, Skyros and Alonissos, however, are more geared to meet the needs of independent travellers. Skiathos has some excellent luxury hotels; most of the accommodation on the other islands is to be found in smaller hotels, guesthouses and apartments.

Skiathos
LUXURY
Aegean Suites Hotel,
Ftelia, tel: 24270 24069,
fax: 24270 24070. This is the newest and most sybaritic luxury hotel on Skiathos. It is the best place to stay if you enjoy being pampered.

MID-RANGE
Kassandra Bay,
Vasilias, tel: 24270 24201,
fax: 24270 21992. An excellent beach hotel, and the amenities are especially good for families – children will enjoy it here.

The Sporades at a Glance

Skopelos
LUXURY
Dionysos,
Skopelos harbour, tel: 24240 23210, fax: 24240 22954. This modern mansion offers excellent accommodation. It has fine sea views and a good swimming pool.

MID-RANGE
Pleoussa Studios,
Livadi, Skopelos town, tel: 24240 23141, fax: 24240 23844. This small and very friendly apartment complex is conveniently situated for a good island holiday – it is close to the town and also the beach.

Skyros
LUXURY
Nefeli, Chora, tel: 22220 91964, fax: 22220 92061. This is a small luxury hotel and the rooms are decorated in distinctive Skyriot style. It offers its guests excellent views and it has its own swimming pool.

Alonissos
Konstantina's Studios,
Old Town, tel/fax: 24240 66165. Nine comfortable studios are available for rent in this traditional house with its super views.

WHERE TO EAT

Skiathos
1901,
Grigirou Pemptou, Agios Nikolaos, tel: 24270 21828. This restaurant offers imaginative Greek and Italian-influenced cooking, and the food is complemented by a good wine list.

Karnagio,
on the harbour, Skiathos town, tel: 24270 22868. This is probably the finest seafood restaurant in Skiathos.

Alonissos
Paraportiani,
Alonissos Old Town, tel: 24240 66165. Seafood taverna and grill.

Skopelos
Alexander Taverna,
Skopelos town, tel: 24240 22324. This is an old-fashioned courtyard taverna with imaginative meat dishes

and grills, and it is situated in the older part of town uphill from the port.

Terpsi,
Panormos Road (4km; 2.5 miles from Skopelos town), tel: 24240 22053. This establishment serves superb roasts and charcoal grills.

SHOPPING

There are many shops in Skopelos: look out for those specializing in handicrafts like ceramics and carved wood, leather, gold jewellery, local sweets and wines, olive oil, clothing and souvenirs.

USEFUL CONTACTS

Tourist Police:
Skiathos, tel: 24270 21111.
Skopelos, tel: 22424 22235.
Skyros, tel: 22220 91274.
Alonissos, tel: 24240 22222.

Websites:
www.greeka.com
www.skopelos-island.com

SPORADES	J	F	M	A	M	J	J	A	S	O	N	D
AVERAGE TEMP. °F	48	48	52	61	70	79	82	81	75	68	57	50
AVERAGE TEMP. °C	9	9	11	16	21	26	28	27	24	20	14	10
HOURS OF SUN DAILY	4	5	6	7	10	11	12	11	9	7	5	4
RAINFALL ins.	2	2	2	1.5	1	0.5	0	0.5	0	1.5	2.5	2
RAINFALL mm	43	45	54	32	21	10	2	9	3	39	62	58
DAYS OF RAINFALL	9	9	8	7	6	3	1	2	2	6	9	9

9
The Ionian Islands

The islands of the Ionian Sea lie in a long chain within sight of the western Greek mainland coast, with the lonely island of **Kythera** far to the south, off Cape Matapan. Unlike the Aegean isles, the Ionian islands were occupied by the Turks only briefly, if at all, and their culture, music and architecture are strongly influenced by the long centuries of Venetian empire. There are faint echoes, too, of a half-forgotten era of British rule during the first half of the 19th century. The Union Jack returned with a vengeance a century later, this time on the shorts and T-shirts of British package holiday-makers. Four decades after the arrival of the first charter jets, much of Corfu's individualism has been eroded by mass tourism, but there is still much to discover here and on its Ionian neighbours.

The Ionians enjoy more winter rain and slightly cooler summers than the southern Aegean isles, and their landscapes are less harsh than their barren sisters in the Cyclades. Olive groves, cypresses, orchards, vineyards and vegetable fields cover much of **Corfu**, **Kefalonia**, and **Zakynthos**, and tiny **Paxi** is almost entirely covered in olives. Like Corfu, Zakynthos and Kefalonia have become popular package holiday destinations thanks to a combination of excellent beaches and airports capable of receiving charter flights. **Paxi** is popular with yacht sailors and the villa crowd and **Lefkada** is a magnet for windsurfers and dinghy sailors, as is **Ithaka**. Between Lefkada, Ithaka and the mainland, an archipelago of tiny islands – most of them uninhabited – provides exciting sailing.

ALBANIA
TURKEY
●ATHENS
MEDITERRANEAN SEA

DON'T MISS

★★★ **Corfu:** the Venetian town with streets which never fail to delight.
★★★ **Zakynthos:** the views of islands and mainland from the summit of Mt Skopos.
★★ **Kefalonia:** walks through the forest on Mt Ainos.
★★ **Lefkada:** the superb beaches of the east coast.
★ **Paxi:** the sea caves on the west coast and Antipaxi islet.

Opposite: *Vlacherna is joined to the Kononi peninsula on Corfu by a causeway and Mouse Island sits just offshore.*

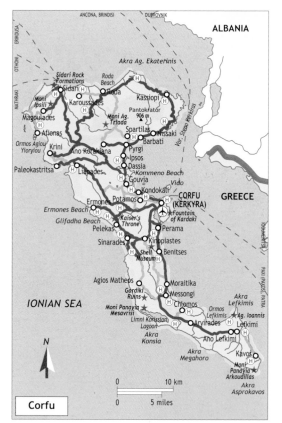

CORFU (KERKYRA) ★★

Corfu is the first taste of Greece for millions of visitors each year and it seems incredible that such a relatively small island (592km²; 229 sq miles) can absorb such numbers. Yet Corfu somehow manages to retain its beauty, and the friendliness of Corfiots is still a matter of local pride – and its sandy beaches are still among the best in Greece.

Kerkyra (Corfu town) is the largest and most attractive town in the Ionian islands, and its old town retains a strong Italian flavour – along with French and British contributions. The Venetians laid the town out in the 14th century, and the narrow streets and tall houses of the **Campiello**

Opposite: *Typical of Corfu town's elegance is Guildford Street with the town hall square.*
Right: *Undoubtedly popular but scenically superb – Paleokastritsa is a favourite destination.*

district would not look out of place on the other side of the Adriatic, with their tiled roofs, wrought-iron balconies and pastel-pink or yellow stucco. The Venetians also endowed the town with two impressive fortresses – the **Paleo Frourio** ('Old Fortress') which was begun in Byzantine times and where a sound-and-light show is held most evenings in summer, and the **Neo Frourio** ('New Fort') built during the 16th century when the Turkish threat was at its highest. During the brief French occupation of the island during the Revolutionary Wars, Parisian-style arcades sprang up around the **Spianada** (Esplanade) and the row of café-arcades called the **Liston** is modelled on the Rue de Rivoli. The British colonial legacy can be seen in a scattering of elegant Georgian buildings, including the **Ionian Academy,** founded by the first British governor, Sir Thomas Maitland, and an incongruous bandstand at the south end of the Spianada.

Museums in the town

Above: *Old buildings with their shuttered windows are a well-preserved feature of Venetian rule.*

include the **Museum of Asiatic Art**, housed in the former **Palace of St Michael and St George** at the north end of the Spianada (open 08:00–14:30 Tue–Sun) and the **Archeological Museum** in Garitsa (open 08:00–15:00 Tue–Sun). The **Byzantine Museum** (open 08:45–15:00 Mon–Sat and 09:30–14:30 Sun) stands on the Campiello, in the heart of the old Venetian quarter and next to the **Orthodox Cathedral**, dating from 1577. Locals pray over a silver reliquary containing the bones of **Agios Spyridon**, Corfu's patron saint, in the church named after him.

Corfu's busiest (and noisiest) resorts are on the east coast, south of the town, at **Benitses**, **Moraitika**, **Messongi** and **Kavos**, with dozens of cheap bars and music clubs that stay open until the early hours. North of Corfu town, tourism development is almost continuous through **Kontokali** and **Gouvia** to **Dassia**, where the large package-tour resorts merge with a long strip of downmarket hotels, bars and tavernas at **Ipsos**, on a narrow shingle beach that is refreshed each summer with sand dredged from the sea bed.

To the north, **Kassiopi** appeals to an older and more up-scale clientele and there are some very exclusive private villas around the coast between **Nissaki** and **Kassiopi** and on the slopes of **Mt Pantokrator**, the island's highest mountain (906m/2800ft). **Roda** and **Sidari** on the north coast are popular with families on package holidays, with gently shelving sandy beaches.

The west coast is less developed, except around dazzlingly picturesque **Paleokastritsa**, where a near-circular turquoise bay with a crescent of beach is overlooked by pine-covered slopes and white limestone cliffs. To the north, **Agios Stefanos** has a long, sandy and occasionally windswept beach, with a scattering of hotels and places to eat and good water sports. To the south of Paleokastritsa, **Agios Gordis** has a fantastic 2km (1.5-mile) sandy beach that has, amazingly, escaped over-development and has become a thriving water-sport and scuba-diving resort.

PAXI (PAXOS) ★★

The smallest of the main Ionian islands (23km²; 9 sq miles), Paxi is almost covered with silver-green olive groves. **Lakka**, at the north end of the island, is favoured by windsurfers and **Longos** is a delightful little fishing village. **Gaios**, the miniature capital, is a favourite yachting port, set on a narrow channel sheltered by an offshore islet on which stands the 15th-century Venetian fortress of **Agios Nikolaos**. Two nearby islets, **Panagia** and tiny **Antipaxi**, the latter with superb fine sand beaches at **Voutoumi** and **Vrika**, are connected by boat; **Mongonissi** is connected by a bridge.

LEFKADA (LEFKAS) ★★

Dominated by barren mountains and linked to the mainland by a road bridge, Lefkada has become a favourite resort for windsurfing enthusiasts and dinghy sailors and the main starting point for yacht flotilla holidays in the Ionian. **Nidri**, midway along the east coast, stands on a sheltered, shallow lagoon that is perfect for windsurfing novices and has become the island's main resort. **Vassiliki**, facing south on a wide turquoise bay, is less developed and has a fine long pebbly beach. There are other beaches in the south at **Poros** and **Dessimi**. **Agios Nikitas**, a small resort on the northwest coast, is more up-market, and boasts some of the island's best restaurants. The rest of Lefkada's rocky west coast is undeveloped, but the spectacular beach at **Porto Katsiki**, beneath its towering limestone cliffs, is a popular day trip by caique.

SEA CAVES

There are seven caves hidden among the sheer limestone cliffs off the west coast of Paxi that can be visited by local boat trips. **Grammatiko** is the biggest, **Ipapanti** is traditionally Homer's wild cove and cave, **Kastanitha** has a roof 185m (607ft) high and **Ortholithos** is guarded by a monolith at its entrance.

A natural 'bridge' (*Tripitos*) at the **Mousmouli Cliffs** is another feature that is visible from the sea.

LEFKADA'S BEACHES

Although the west coast is rugged some superb beaches nestle beneath the cliffs at **Pevkoulia**, **Kathisma** and **Kalamitsi**. At the southern tip lies **Kavo tis Kyras** (lady's cape) from whence a distraught Sappho, rejected by Phaon, is supposed to have thrown herself. Nearby **Vassiliki** is regarded as heaven by windsurfers.

ODYSSEUS'S FOOTSTEPS

The islanders of **Ithaka** have made great play of their links, however tenuous or fanciful, with Odysseus, son of Laertes and Anticleia. West of Vathi at **Marmarospilia** is the Cave of the Nymphs and at **Dexia** a sleeping Odysseus was carried ashore by the Phaeacians. Disguised as a beggar Odysseus met swineherd Eumaeus at **Elliniko**.

North of Vathi at **Aetos** is the so-called Castle of Odysseus; at Agros is the Field of Laertes.

Opposite: *Villagers in Kefalonia have a love affair with flowers; the cottages near Marko reflect this.*

Lefkada town was hit hard by earthquakes in 1948 and 1953 and few interesting buildings have survived except for the old Frankish castle of **Agia Mavra** (Santa Maura) and the 18th-century churches of **Agios Minas**, **Agios Dimitrios**, **Agios Spyridon** and **Pantokrator**. The **Orpheus Folklore Museum** (open daily 10:00–13:00 and 18:00–21:00) has a display of exquisite local embroidery and old maps and more embroidery is displayed at the **Maria Koutsochero Museum** in Karya. Fine needlework is a Lefkada tradition and beautiful emroidery is obtainable in **Karya** and other mountain villages including **Englouvi**.

Situated offshore from Nidri is an archipelago of tiny islands including **Skorpios** (the private retreat of the multimillionaire Onassis shipping dynasty), **Mandouri**, **Sparti** and **Skorpidi**. The largest, **Meganisi** (23km²; 9 sq miles) is a popular day excursion, with excellent fish tavernas at its port, **Vathi**, but accommodation is limited. A sea cave, the **Papanikolaos Grotto**, is included on the excursion route. The **Festival of Agios Konstantinos** brings the crowds to the island on 21 May.

ITHAKA (ITHAKI) ★★

Ithaka was the island kingdom of Homer's cunning hero Odysseus (Ulysses). Today, this mountainous isle with its deep blue bays and wooded hinterland attracts a following looking for peace and quiet.

Vathi, the capital, is the prettiest small town in the Ionians, with dignified mansions and neoclassical buildings (all restored after the 'quake of 1953) set around a wide bay full of yachts and fishing boats. Two ruined forts (Loutsa and Kastro) are set either side of the harbour entrance. From the harbour there are caique services to good beaches at **Bimata**, **Filiatro**, **Skinos** and **Sarakiniko**. An icon of Christ in the **Church of the Taxiarchis** is claimed to be the work of the young El Greco. The **Archeological Museum** has a local collection of Classical and Mycenaean objects (open 08:30–14:30 Tue–Sun).

KEFALONIA CAVES

Sami has become the main
ferry port of Kefalonia and a
growing resort – the attraction
is the marvellous caves.
Melissani cave, once a sanctuary of the God Pau, has an
indoor lake dappled in shades
of blue as filtered light from a
hole in the roof plays on it.
Near the village of Chaliotata,
the **Drogkarati cave** is big
enough to hold concerts in,
with a brightly lit wonderland
of stalactites and stalagmites.

KEFALONIA ★★★

'Captain Corelli's Island' has become a smash hit with
(mainly British) package holidaymakers since the 1990s
and features in virtually every holiday brochure, but this
lush and beautiful island is still far from overdeveloped.
Kefalonia has no spectacular ancient sites, but in Classical
times there were four city
states on the island (Sami,
Pali, Krani and Pronnoi).
The scant remains of **Krani**
can be seen southeast
of the island capital,
Argostoli. In Byzantine
times Kefalonia was raided
by Arab corsairs and
Normans from Sicily be-
fore falling to Venice. It
was seized by the Turks
from 1483–1504 before
reverting to the Venetians.
Two dramatic hilltop
castles, one set above the
picturesque little resort of
Assos with its pretty
harbour, the other on a
hilltop at **Agios Georgios**
(Kastro), are a Venetian
legacy.

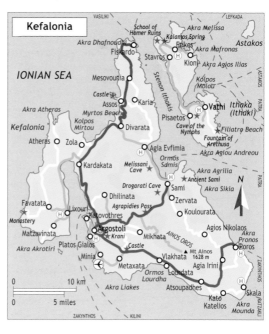

Kefalonia

Right: *Idyllic Fiskardo on Kefalonia has become a favourite with yachtsmen exploring the islands.*

Kefalonia's natural attractions include two spectacular caves – **Drogarati**, a huge cavern filled with strange rock formations, and **Melissani**, with a luminous, deep blue lake in an echoing grotto which was once sacred to the god Pan.

Argostoli, the capital, is a drab commercial town set on a bleak grey lagoon which is traversed by the remarkable **Drapanos Bridge**, a causeway built during the British era. Levelled by the 1953 earthquake, it has benefited from a recent makeover, giving it a lively main pedestrian shopping street lined with cafés. The town's **Archaeological Museum** (open 08:30–15:00 Tue–Sun) has finds from ancient sites including a fine bronze head from ancient Sami and the **Historical and Folk Museum** (open 09:00–14:00 Mon–Sat) has a collection of household items, farm tools and island costumes.

Kefalonia's most popular beach resorts are on the **Lassi peninsula** (west of the capital) and along the south coast, east of **Argostoli**. **Skala** and **Lourdata** are among the best beaches. An impressive lighthouse

(Agios Theodori) sits at the peninsula's tip. On the northern peninsula, there is a superb beach beneath awesome cliffs at **Myrtos** (North of Asos) and the lovely village of **Fiskardo** has become the island's premier resort, with a quayside esplanade lined with cafés and restaurants, colourful old stone houses, and a harbour full of yachts and motor cruisers. Fiskardo has small beaches nearby,

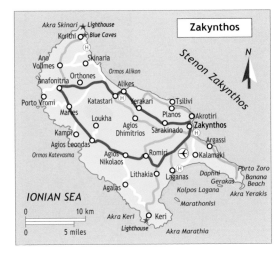

and motorboats can be hired for swimming and snorkelling trips to nearby bays.

Poros, near the southern tip of the island, is a sleepy port town that exists only to handle ferries from the mainland. **Sami**, the island's main port, is midway along the east coast, with a long waterfront and a few small hotels. **Agia Evfimia**, on the north shore of Sami Bay, has developd into a cheerful small resort town with a pebbly beach, scuba diving and excursions by boat to Fiskardo and Ithaca.

ZAKYNTHOS (ZANTE) ★

This fertile island with some excellent sandy beaches has been almost overwhelmed by package tourism. Severely damaged by an earthquake in 1953, its towns and villages were completely rebuilt and there are few historic monuments or buildings of note. The island's biggest draw is the huge sandy beach at **Laganas Bay**, where tourism development threatens to destroy one of the last nesting places of the endangered loggerhead turtle. Other beaches are at **Argassi**, **Tsilivi**, **Planos** and **Alikes** on the east coast, all of which are thoroughly developed for tourism.

ZAKYNTHOS'S BEACHES

Loggerhead turtles (*see p. 11*) were first to recognize the clear water and fine expanse of sand of **Laganas** long before beach umbrellas, cafés, discos or tavernas. It is possible to escape the crowds at **Porto Zoro** and **Banana Beach** where the sands are full of white sea daffodils in high summer. Further towards the peninsula tip are excellent beaches at **Gerakas**, **Daphni** and **Sekanika**.

The tiny offshore islet of **Marathonissi** has a good beach and is popular with day-trippers.

Right: *Bustling Zakynthos town with its street cafés shows no traces of the earthquake which decimated it in 1953.*

Zakynthos town has a waterfront parade dominated by mock-neoclassical buildings, including the **Byzantine Museum**, with a collection of religious art rescued from churches demolished by the 'quake, including paintings by the famed Cretan icon-painter Michael Damaskinos. The ruins of the **Venetian Kastro** (fortress) are reached by a cobbled path from the town square (open daily 08:00–20:00). Zakynthos's cliff-lined, inaccessible west coast is devoid of tourism, but daily boat cruises visit the 'blue cave' at the island's north tip and the gradually disappearing 'Smuggler's Wreck' on a dazzling, empty white each near **Porto Vromi**.

From the white church of **Panagia Skopiotissa** on the summit of **Mt Skopos**, dominating the Gerakas peninsula, there are superb views over the island and the Peloponnese mainland.

KYTHERA ★★

Kythera is the joker in the Ionian pack. Remote from its siblings, its cliff-lined coasts rise from the **Kythera Channel** between Crete and Cape Maleas. Remote and unspoiled, Kythera (area 278km²; 107 sq miles) is claimed to be the birthplace of Aphrodite, and the ovoid islet of **Avgo** nearby is said to be the egg from which she hatched. Kythera's population plummeted from around 40,000 at the end of the 19th century, when many islanders emigrated to Australia, to less than 5000 today. Many Greek-Australians return to their ancestral isle in summer.

Chora, the capital, is a breathtakingly pretty village of whitewashed houses set high above the port of **Kapsali**, with its twin pebbly bays on the island). Venetian coats of arms can be seen on some of the old mansions in Chora, and the Venetian castle is a romantic ruin littered with old cannon left over from the successive Venetian, Russian and British occupations of the island. Strikingly pretty villages are dotted around the highhy plateau that is the island's hinterland, among them **Paleochora**, the hidden Byzantine-era capital which was sacked by the Turkish corsair Khair ed Din (Barbarossa) in 1537, and **Mylopotamos**, known for its small streams and watermills.

Kato Chora is a semi-deserted fortified village atop a steep cliff, with the Lion of St Mark, symbol of Venice, carved above its castle gateway. There are more beaches around 8km (5 miles) east of **Kapsali** at **Fyri Ammos** and **Chalkos**.

In the northern part of the island, **Potamos** is an attractive market town with some good small hotels. **Agia Pelagia**, the main ferry port, is little more than a harbour with a few hotels attached.

Between Kythera and Crete, tiny **Antikythera** is a rocky islet with two tiny hamlets and a once weekly ferry: ideal for solitude seekers.

ELAFONISSOS

Only a few hundred metres separates Elafonissos from Neapolis on the southern mainland. It has one small fishing village, facing the mainland, with a few tavernas and guesthouses, but its well-kept secret is a huge sweep of fine sand on an isthmus closwe to its southern tip. There are no facilities here, but the beach attracts a gaggle of bohemian campers each summer.

Kythera

Platia Ammos
Agia Pelagia
Potamos
Paleochora
Arondika
Diakofti
Agia Sophia Cave
Temple of Aphrodite
Kato Chora
Avlemonas
Mylopotamos
Mirtidion Monastery
Fratsia
Paliopolis
Fyri Ammos Beach
Chora
Kapsali
0   5 km
0   3 miles
Chalkos Beach

THE ANTIKYTHERA MECHANISM

In the diving world, Antikythera is renowned for its shipwrecks, especially through discovery in 1900 of the Antikythera Mechanism. Now in the Archaeological Museum in Athens, this device is an intricately geared astronomical computer made in Rhodes in 82BC, which showed the phases of the moons and planets.

The Ionian Islands at a Glance

Winters can be cold and damp with heavy storms. Spring is flower season in Zakynthos, Corfu and Kefalonia. Crowds arrive Jun–Sep. Oct is crowd free and the sea is warm.

GB Airways (www.gbairways. com) operating on behalf of British Airways (www.ba.com) operates scheduled **flights** between London and Corfu (summer only). There are many summer charter flights from the UK, Ireland, Belgium, Germany and the Netherlands to Corfu, Zante, and Kefalonia and to Preveza on the mainland for Lefkada. Trans-Adriatic **ferries** from Italian ports including Ancona, Bari and Brindisi call at Corfu, Igoumenitsa, Patras and Kefalonia (summer only) and international ferries go from Igoumenitsa and Patras to Albania and Croatia. Short-haul ferries link the Ionian islands with each other and with ports on the western mainland: Igoumenitsa (for Corfu and Paxi) and Patras (for Corfu, Ithaka, Kefalonia and Zakynthos). Ferries go to Ithaka and Kefalonia from the small mainland port of Astakos, to Kefalonia, Zakynthos and Ithaka from Killini, south of Patras, and to Paxi and Corfu from Sivota and Parga, but the smaller ports are less convenient for onward connections. The Ionian isles are near the mainland; most crossings take 1–3 hours. In summer, **hydrofoils** connect Kythera with

Piraeus, the Argo-Saronic isles, and the mainland port of Monemvasia. Ferries also connect Kythera and Antikythera with Gythion, on the mainland, and Kastelli, on Crete. For timetables, visit www.gtp.gr
International ferry lines: Blue Star Ferries, tel: 1089 19950, or (Corfu) tel: 26610 81222; (Kefalonia) tel: 26740 22055; (Ithaka) tel: 26740 33120. Maritime Way, (Corfu) tel: 26610 25000, (Kefalonia) tel: 26740 23007. Minoan Lines, (Corfu) tel: 26610 25000.
Local ferry lines: Blue Star Ferries (see above): Patras to all main Ionian islands. Kerkyra-Igoumenitsa Lines, tel: 26610 22275. Kerkyra Lines, tel: 26610 21991. Miras Ferries, tel: 21041 74459 (Zakynthos to Killini). Neapolis-Elafonissos Lines, tel: 27340 61177. Zakynthos Lines, tel: 26950 44117. Zantefast Ferries, tel: 26950 49500.
KTEL long-haul **buses** connect the Ionian ports with Athens, arriving/departing at Kifissiou 100 (A) Terminal (tel: 1440 for schedules). Individual island bus and ferry connections to Athens: tel: 21051 29443 (Corfu); 21051 50108 (Lefkada); 21051 50785 (Kefalonia); 21051 29432 (Zakynthos).
Port authorities:
Patras, tel: 26103 41002.
Igoumenitsa, tel: 26650 22235.
Corfu, tel: 26610 23277.
Paxi, tel: 26620 32259.
Lefkada, tel: 26450 22322.
Kefalonia (Sami), tel: 26740.
Zakynthos, tel: 26950 28117.

Bus services on Zakynthos and Corfu are regular but are more limited on the smaller islands with no more than one or two services a day linking towns on Lefkada, Kefalonia and Kythera. On all islands **taxis** are cheap and comfortable way of making trips; remote beaches can be reached by **caique** service in season (see p. 124 for on-line timetable and booking). Small **motorboats** can be hired from several resorts, including Kefalos (Corfu), Nidri (Lefkada) and Fiskardo, Assos and Agia Evfimia (Kefalonia).

Accommodation can be hard to find on Lefkada, Kefalonia, Ithaka and Zakynthos. Corfu is better for independents, but virtually all accommodation on Paxi is in villas and apartments; booking essential. Kythera is busy with local visitors Jul–Aug.
Corfu
LUXURY
Corfu Palace, Dimokratias 2, tel: 26610 39485, fax: 26610 31749. Impressive hotel with a pool and rooms with sea views near the historic town centre.
Pelekas Country Club, tel: 21066 40077, fax: 26610 52919. Studios and suites in a Venetian country estate; pool.
MID-RANGE
Fundana Villas, Paleokastritsa, tel: 26630 22532, fax: 26630 22453. Self-catering apartments in old Venetian mansion, 4km (2.5 miles) from Paleokastritsa.

The Ionian Islands at a Glance

Paxi

MID-RANGE

Paxos Beach, Gaios, tel: 26620 32211, fax: 26620 32695. Bungalows near pebble beach.

Lefkada

MID-RANGE

Red Tower, Nidri, tel: 26450 92951, fax: 26540 92852. Rooms and apartments in modern complex with pool.

BUDGET

Nefeli Hotel, Agios Nikitas, tel: 26450 97400, fax: 26450 97401. Comfortable small hotel near Agios Nikkitas's beach.

Kefalonia

LUXURY

Emelisse, Emelissos beach, Fiskardo, tel: 26740 41200, www.arthotel.gr By far the most stylish hotel on Kefalonia, with pool and room service.

MID-RANGE

Aignantia, Thesi Tselentata, Fiskardo, tel: 26740 51801, www.agnatia.com This small and friendly hotel with just 16 rooms offers value for money in Kefalonia's prettiest village. **Porto Skala Hotel Village**, Skala, Agios Giorghios, tel: 26710 83501, www.porto skala.com Cosy, small-scale resort on the south coast.

Kythera

MID-RANGE

Margarita, Chora, tel: 27360 31711, fax: 27360 31325. Pretty rooms in immaculately restored neoclassical mansion in whitewashed village. **Raikos, Kapsali**, tel: 27360 31629, fax: 27360 31801. Bungalow rooms, great views; only pool on the island.

Zakynthos

MID-RANGE

Castelli Hotel, Agios Sostis, Laganas, tel: 26950 52367, www.castellihotel.com Well priced three-star with pool, broadband Internet access.

BUDGET

Porto Zorro, Vassilikos, tel: 26950 35304, www.porto zorro.gr Near one of the better beaches, this might be the best value on Zakynthos. **Hotel Vanessa**, Kalamaki beach, tel: 26950 26713. In May and Oct it offers some of the cheapest rates on the island, but with excellent facilities.

WHERE TO EAT

Corfu

Rex, Kapodistriou 66, Corfu town, tel: 26610 39649. Old-fashioned Corfiot specialties.

Paxi

Vasilis, Longos, tel: 26620 31587. Much-loved taverna; seafood and grilled meat.

Lefkada

Sappho, Agios Nikitas, tel: 26450 97497. Good fish restaurant on Agios Nikitas beach.

Kefalonia

Tassia, Fiskardo harbour, tel: 26740 41205. One of the best in the islands; a sophisticated menu and price tag to match.

TOURS AND EXCURSIONS

Corfu: day trips to Antipaxi, Paxi, and Parga and Sivota on the mainland; escorted coach tours (2–3 days) of northern mainland via Igoumenitsa. Day trips to Albania are possible. **Paxos**: boat trips to Antipaxi, Parga. **Lefkada**: boat trips from Vasiliki to Fiskardo (Kefalonia) and Ithaka; boat trips from Nidri to Meganissi and around Skorpios. **Kefalonia**: boat trips to Vasiliki and Ithaka. **Zakynthos**: boat trips to 'Blue Grotto' and 'Smugglers' Wreck'. **Kefalonia** and **Zakynthos** (via Killini on the mainland): ancient Olympia (an hour's drive), and the Frankish medieval castle of Chlemoutsi, 5km (3 miles) from Killini harbour.

USEFUL CONTACTS

Ionian Islands Tourism Directorate, Alykes, Corfu, tel: 26610 37693. **Kefalonia Tourism Office**, Provlita Teloniou, Argostoli, tel: 26710 22248.
Tourist Police:
Corfu, tel: 26610 30265.
Ithaka, tel: 26740 32205.
Kefalonia, tel: 26710 22200.
Lefkada, tel: 26450 31218.
Kythera, tel: 27350 31206.
Zakynthos, tel: 26950 24450.

IONIAN ISLANDS	J	F	M	A	M	J	J	A	S	O	N	D
AVERAGE TEMP. °F	57	57	61	64	73	81	84	84	81	73	66	61
AVERAGE TEMP. °C	14	14	16	18	23	27	29	29	27	23	19	16
RAINFALL ins.	4.5	4.5	3	2.5	1	0.5	0	0.5	1.5	4	6	5
RAINFALL mm	111	113	76	62	20	13	6	9	35	105	152	127
HOURS OF SUN DAILY	4	5	6	7	10	11	12	11	9	7	5	4
DAYS OF RAINFALL	13	13	10	9	5	1	1	2	4	10	12	14

Travel Tips

Tourist Information

The **Hellenic Tourism Organisation** (EOT) produces an excellent range of free maps and brochures for all islands and island groups, and also provides free ferry timetables and hotel listings. **London**, 4 Conduit Street, tel: 020 495 9300, fax 0207 287 1369; **New York**, Olympic Tower, 645 5th Ave, Suite 903, tel: 212 421 5777, fax 212 8266940; **Australia/NZ**, 37-49 Pitt Street, Sydney, tel: 924 11663, fax 924 12499; **Canada**, 1500 Don Mills Road, Suite 102, Toronto, tel: 968 2220, fax 968 6533; **Athens**, Leoforos Amalias 26, Syntagma, tel: 21033 103912/01716/10640; and Elevtherios Venizelos International Airport arrivals hall, tel: 21035 45101/21035 30445 (daily 08:00–22:00).

Entry Requirements

From 1995 EU citizens can stay indefinitely; most other visitors (including Australia, Canada, New Zealand and the USA) are allowed up to three months (South Africa: two months), no visas required. Children must either hold their own passports or be entered in parental passports. Visa extensions or work permits can be obtained from the Aliens Bureau, 9 Halkokondili Street, Athens, tel: (01) 362-8301. Apply well in advance as the bureaucracy is time-consuming. Temporary jobs including bar and restaurant work are not difficult to find on the islands – ask in the local *kafeneíon*. Graduates hoping to find jobs teaching English will need a TEFL qualification.

Customs

Visitors from EU countries may bring in unlimited quantities of cigarettes, cigars, tobacco, wines and spirits, perfumes and other goods provided that they are for personal consumption or gifts and are not to be resold. In theory, only one camera is allowed per person and articles such as lap-top computers and windsurfers call for assurance from a Greek national resident in Greece that they will be re-exported: in practice there are few problems and foreign nationals are not often stopped.

Health Requirements

No certificates of vaccination are required for visitors.

Getting There

By Air: Direct scheduled flights operate daily to Athens from London and New York, less frequently from Montreal, Toronto, Sydney and Melbourne. Airlines flying from the UK include British Airways, EasyJet, and Olympic Airlines. There are also direct scheduled flights to Thessaloniki in northern Greece. In summer, British Airways also flies several times weekly from London (Gatwick) to Heraklion and Rhodes. There are charter flights in summer from most UK and mainland European airports to island airports on Crete, Corfu, Kefalonia. Kos, Lesbos, Mykonos, Samos, Santorini, Skiathos, Rhodes and Zakinthos. These may be bought as part of a flight and accommodation package, or as flight-only. Aegean Airlines and the domestic wing of Olympic Airlines fly to many of the islands, including all those listed above, plus Karpathos, Kithira, Ikaria, Chios, Milos and Kastellorizo. Charter passengers may only visit Turkey for a day and may not stay overnight; if your passport indicates a longer stay in Turkey the return section of your charter ticket from Greece may be invalidated.

By Road: Strife in the former Yugoslavia made the route through Zagreb, Belgrade and Skopje to northern Greece problematic. The threat of violence has receded, and buses now take this route to Thessaloniki. Those taking their own car to Greece are better advised to drive to Ancona in northern Italy and travel by ferry to Corfu, or to Igoumenitsa or Patras on the mainland. The bus journey from London to Greece takes 56–66 hours via Brussels or Frankfurt Eurolines, tel: (UK) 08705 143 219, (Ireland) 0183 66111, www.eurolines.com

By Rail: There are no longer direct trains from London to Athens but you can go through Italy via Paris and Bologna to Brindisi and take the ferry to Patras. Rail Europe: www.raileurope.co.uk

By Boat: Regular boats go from Brindisi to Corfu, Igoumenitsa, Patras and less frequently from other Italian ports: Ancona, Bari and Otranto. To reach Athens either take a ferry from Corfu and then a bus, or a ferry to Patras with connections by bus and train to Athens. Other ferries run to Piraeus from Alexandria, Haifa, Limassol, Syria, Istanbul and less frequently from Odessa. From Piraeus and several other ports, a number of ferries and hydrofoils connect with all the inhabited islands. Timetables change annually and the best source of information is the Athens Gazette (on sale in all bookshops and kiosks in Athens). Details pertaining to each island are given in the relevant At a Glance. In general, hydrofoils are twice as fast as ferries but double the price.

What to Pack

In summer, lightweight cotton T-shirts and shorts suffice most of the time. Out of season evenings can be cool so pack a pullover (even waterproofs). In more up-market resorts you may want smarter clothes for evening but leave the tuxedo at home. Hats, sunglasses and a UV protection sun cream are advisable during the day.

USEFUL PHRASES	
ENGLISH	**GREEK**
yes	*né*
no	*ochi*
hello	*khérete*
how are you?	*ti kánete?*
goodbye	*adio*
please	*parakaló*
thank you	*efkharistó*
sorry/excuse me	*signómi*
how much is?	*póso iné?*
when?	*poté?*
where?	*pou?*
I'd like	*thélo*
open	*aniktó*
closed	*kleistó*
one	*éna*
two	*dhío*
three	*tría*
four	*téssera*
five	*pénte*
six	*éxi*
seven	*eftá*
eight	*okhtó*
nine	*enniá*
ten	*dhéka*

Money Matters

Currency: Greece adopted the euro in 2001. There are 5, 10, 20, 50 and 100 euro notes and coins of 1 and 2 euros and 1, 2, 5, 10, 20 and 50 cents.

Currency exchange: other than on the remotest islands (where you take euros with you) you can change money in a bank (*trápeza*), post office or shipping agent. Post Offices change cash, travellers cheques and eurocheques and charge less commission than banks. In major resorts there are many ACT's (automatic cash tellers).

Travellers cheques: especially Thomas Cook and American Express are accepted in all banks and post offices (passport needed as ID). Cash transfers

are best handled by major banks in Athens or Piraeus.

Credit cards: allow cash withdrawals at banks and ACT's. Visa is handled by the Commercial bank of Greece and Access/Mastercard by the National Bank of Greece.

Tipping: a 10–15% service charge is added to restaurant bills, but Greeks leave change as a tip. Taxi drivers, porters and cleaners welcome tips.

VAT: The basic rate of VAT is 19% (accommodation, bars and restaurants) but there are lower rates for books and newspapers (4.5%) and food, medicines and other essential goods (9%). Non-EU travellers can claim back the VAT paid on most purchases when they leave, if they have the original receipts. Many tourist shops are part of a scheme that helps to fast-track VAT reclaims for visitors – look for the prominent VAT reclaim scheme logo.

Accommodation

From June to early September most island hotels are geared to the pre-booked package trade – look for last-minute bargains from your local travel agents. Prices are government controlled according to category (Luxury, A, B, C, D and E). By law, these rates have to be displayed in each hotel room. On smaller islands many hotels close out of season but people are glad to rent rooms and are open to gentle bargaining. The Tourist Police and NTOG office have lists of available accommodation (including pensions) on any island and locals stand on the harbour to meet ferries and offer rooms for rent. Greek ladies are house proud and the accommodation will be simple but spotless.

Youth hostels: both official (curfew) and informal (no curfew) exist on many islands – an International Membership Card is available from the Greek Association of Youth Hostels, 4 Dragatsaniou Street, Athens, tel: 21032 34107.

Camping: is offically permitted only on authorized sites – on smaller islands be guided by local attitudes over sleeping out on beaches. If police insist you move, be polite and do it. Very little provision is made for disabled visitors and their companions. Visitors with disabilities should contact the NTOG. See www.gtp.gr and www.greekislands.gr for links to hotels, pensions, apartments and guesthouses throughout the islands and mainland.

Eating Out

Most visitors eat in restaurants (*estiatoria*) or tavernas. In the latter diners usually begin with starters (*mezédhes*) and follow with meat and fish courses.

Some tavernas specialize in fish (*psarotaverna*) or grills (*psistaria*). Greeks eat late and do not hurry a meal. Serious foodies should invest 19 euros in the *Alpha Guide* (Desmi Publications, Mesogeion 2–4, Athens 11527) to more than 2000 of the best places to eat and stay, updated annually and sold at English newsagents.

Transport

Boat: inter-island services are operated by ferry and hydrofoil (rough seas can play havoc with hydrofoil schedules). To many smaller islands there are caique services, and taxi boats operate between ports/resorts and other beaches. Timetables can be consulted and bookings made quickly and easily via the internet. The centralized service is accessed via www.greekferries.gr which allows you to look at local and international services. Domestic lines include Anev, Minoan, Poseidon, Strintzis and Superfast. Agapitos, which operates in the Cyclades, can be accessed directly via http://agapitos-ferries.com

Air: There are very few inter-island air links. Almost all Olympic and Aegean Airlines flights connect via Athens. Crete-based Sky Express (www.skyexpress.com) flies from Heraklion to Rhodes, Mytilini, Kos, Samos, Santorini and Ikarioa, and links Santorini with Mykonos and Rhodes. Airsea Lines (www.airsealines.com) flies seaplanes from its Patras base, linking the Ionian islands of Paxos, Corfu, Ithaka, Lefkas and Kefalonia with each other and with Patras and Ioannina on the mainland, and has plans to expand its inter-island services to the Cyclades.

Road: on the more popular islands local buses are a reliable mode of transport – they make stops for major beaches. On smaller islands buses might be one a day between towns – leaving early and returning at the end of the working day.

Taxis: are a widely used on all the islands – agree on a price before the journey or check the meter is running. Sharing is common practice – each person pays full rate for the part of the journey they undertake.

Car hire: charges and fuel costs are high in Greece and worthwhile only on larger islands as part of a 'fly-drive'. Non-EC citizens need an International Driving Licence. Always pay the supplement for collison damage waiver to avoid potential problems later. Assistance can be sought via the **Automobile and Touring Club of Greece** (ELPA) at – Athens: 2–4 Messoghion Ave, or 6 Amerikis and Panepistimiou Street; Crete: Hania or Irakliou.

CONVERSION CHART		
FROM	**TO**	**MULTIPLY BY**
Millimetres	Inches	0.0394
Metres	Yards	1.0936
Metres	Feet	3.281
Kilometres	Miles	0.6214
Square kilometres	Square miles	0.386
Hectares	Acres	2.471
Litres	Pints	1.760
Kilograms	Pounds	2.205
Tonnes	Tons	0.984

To convert Celsius to Fahrenheit: x 9 ÷ 5 + 32

PUBLIC HOLIDAYS

Many public holidays centre around religious events and precise times depend on the Orthodox calendar:

1 January • New Year's Day *Protochroniá*

6 January • Epiphany *Epifánia*

February – March • 'Clean Monday' *Katharí Deftéra* (precedes Shrove Tuesday)

25 March • Greek Independence Day *Evangelismós*

Late March – April • Good Friday *Megáli Paraskeví*

Easter Sunday • *Páscha*

Easter Monday • *Theftéra tou Páscha*

May 1 • Labour Day *Protomayá*

15 August • Assumption of the Virgin *Koísmisis tis Theotókou*

28 October • Greek National Day *Ochi Day*

25 December • Christmas Day *Christoúyena*

26 December • Gathering of the Virgin *Sináxi Theotókou*

Helpline tel: 104/105 (English speaking operators). Maximum speed in built up areas is 50kph (30mph) and in other areas 80kph (50mph). Diesel, super and unleaded petrol are available on all the major islands. Car-hire outlets for Avis, Budget and Hertz operate at major airports and in Athens along Singrou Ave. If you take your car to Greece a year's free use is permitted by customs (with a 4 month extension on request) – Australians and North American citizens are allowed two years.

Motorcycles, bicycles and scooters: are cheap and fun for getting around the smaller islands. Check the bike carefully first (especially brakes) and wear protective clothing – accidents are common.

Hitch hiking: is generally safe and accepted, if not speedy.

Tours: specialized tours are geared towards antiquities and wildlife. A list of all operators is provided by the NTOG.

Business Hours

Vary seasonally and from year to year; for up to date opening times and entrance fees for museums and archaeological sites, consult the Ministry of Culture website: www.culture.gr **Banking hours:** 08:00–14:00 Mon–Thu, 08:00–13.00 Sat. Go early; queues can be long.

Time Difference

Greece is two hours ahead of Greenwich Mean Time, one hour ahead of Central European Time and seven hours ahead of US Standard Winter Time. Clocks go forward one hour on the last weekend in March, and back on the last Sunday in September.

Communications

Post: there are post offices (*tachidromío*) and money-changing facilities in large and small towns and at all ports, providing normal postal services (including *poste restante*). English is often spoken. Post can be slow (up to three weeks for a card) and express, though more expensive, is much faster. Stamps (*grammatósima*) are also sold at kiosks and tourist shops. Post boxes are yellow.

Telephones: international calls can be made from hotels or more cheaply from offices of the OTE (Organsimós Telefikoinonía Elládos) – there is at least one on every island allowing you to dial direct or make collect calls. For the UK dial 0044 and then the area code (omitting the zero at the beginning); for the US dial 001. Cardphones have replaced payphones everywhere – phonecards are sold at *periptera*. Greeks have enthusiastically adopted the mobile phone, and cellphone coverage is comprehensive even on small and remote islands. Various options are available to reduce the cost of mobile phone use away from your home country.

Electricity

Mains voltage is 220AC, at 50Hz. Plugs are continental 2-pin – use universal adaptors. US appliances need converters.

Weights and Measures

The metric system is used plus, occasionally, measures from the early Ottoman occupation, such as the *oká* (1.3kg or 2.8lb) divided into 400 *drams* and the *strémma* (0.25 acre).

Health Precautions

Visitors to the islands should make sure that their tetanus protection is up to date. Don't underestimate the strength of the sun – even short exposures to sensitive skins can leave a child or adult very sunburnt and in agony. Remember to use sun hats, high-protection

GOOD READING

Classics (Penguin Classic)
• Homer: *The Odyssey* and *The Illiad*.
• Herodotus: *The Histories*.
• Pausanias: *The Guide to Greece* (2 vols).
• Plutarch; *The Age of Alexander*, *Plutarch on Sparta*, *The Rise and Fall of Athens*.
• Thucydides: *History of the Peloponnese War*.
• Xenophon: *The History of My Times*.
• Durrell, Gerald: *My Family and other Animals*, Viking/ Penguin.
• Kazantzakis, Nikos: *Zorba the Greek*.
• De Bernieres, Louis: *Captain Corelli's Mandolin*. Romance set in Kefalonia during World War II.
• Johnston, Paul: *A Deeper Shade of Blue* (Hodder & Stoughton 2002). Rocking

debut for half-Greek, half-Scots maverick private detective Alex Mavros.
• Leigh Fermour, Patrick: *Mani*.
• Leigh Fermour, Patrick: *Roumeli*. Eclectic musings by the greatest philhellene raconteur of the 20th century.
• Mazower, Mark: *Hitler's Greece*. Sheds new light on the Italian and German occupation of islands and mainland 1941–45.
• Wheeler, Sara: *An Island Apart*. Travelogue through Evia.
• Maclean, Alistair: *The Guns of Navarone*. World War II adventure, loosely based on real events (the evacuation of Allied troops from Samos and Leros in 1943) and filmed (starring Anthony Quinn, David Niven) on location in Rhodes and Kythira.

sun cream and practice sensible sunbathing. The only truly venomous snake is the viper (*kufi*) – anti-serum available locally; large black Montpellier snakes are harmless to humans. At the seaside, weaver fish can lie hidden beneath sand in shallow water with their poisonous spines protruding. Excessive olive oil can cause stomach upsets – retsina, Coca-cola and fresh parsley (*mitanós*) can all help. Although standards of hygiene are generally high in Greece, water shortages (and poor Greek plumbing) on islands don't make things easy in summer. Always carry your own toilet paper. Tap water is safe to drink but may be brackish on some islands – bottled water is widely available.

Health Services

There is a reciprocal agreement giving free medical treatment to EC residents; UK visitors should take the European Health Insurance Card (EHIC). Your travel insurance should have medical cover, including emergency medical evacuation by air. Many doctors speak English but equipment in some hospitals is often behind what is commonplace at home.

Personal Safety

Theft is still not commonplace on Greek islands although it has increased in tourist destinations like Ios in recent years. In cities, Greeks blame the marked increase in car theft and stealing from rooms (now no different from other Euro-

pean cities) on an influx of poverty-stricken Albanians. Generally harassment of lone females is low-key for a Mediterranean country outside the resorts and many women explore the islands alone – a sharp *afístime* (leave me alone) or *fíyete* (go away) usually suffices. Greek friends might equip you with a few more forceful phrases. Much has been made in the western press about incidents of rape – they horrify Greeks as much as anyone because crime is so rare here.

Emergencies

Ambulance: tel: 166 anywhere in Greece.
Police: *see* tourist police under Useful Contacts in every At a Glance section.

Etiquette

In monasteries and churches, remember to dress with decorum: no shorts or bare tops for men; women should cover their bare arms or legs so as not to cause offence or be offended when admission to a place of worship is refused. Nudism is forbidden by law except in designated areas – however, topless sunbathing is permitted on most beaches.

Language

Greek is the main language of daily conversation, notices in shops and on signposts on the islands. Ferry destinations at ports are in Greek capitals. English is learned at school and many Greeks speak a little German. A few words of Greek on your part are welcomed.